Praise for
Climate—A New Story

"There is nothing 'safe' in these writings; almost every chapter courts controversy. We as readers are the beneficiaries of this bravery. This is a message that must be heard loud and clear as we chart a path toward social and ecological renewal."

—HELENA NORBERG-HODGE, author and filmmaker of *Ancient Futures* and *The Economics of Happiness*

"This is a groundbreaking book. Eisenstein makes an inspiring, positive, and convincing case for a full and proper understanding of the present human predicament—a radical shift from a utilitarian worldview to an integral world view rooted in a sense of the sacred which recognizes the intrinsic value of nature and life."

—SATISH KUMAR, founder of Schumacher College and editor emeritus of *Resurgence & Ecologist*

"This book is brave enough, vulnerable enough, insightful enough to activate a truth buried deep within all of our hearts: that the planetary crisis we face today can only be transformed by a revolution of love. It calls each of us to break with our patterns of war thinking and realize our interconnectedness with all life on Earth."

—JODIE EVANS, cofounder of Code Pink

"A clarion call to reconnect through love with our living Earth. Eisenstein offers a deeply analyzed and compelling case to collectively move past divisive reductionism, betwixt false Prophets of doom and false Profits of denial, towards a revitalization of reverential relations."

—BROCK DOLMAN, Occidental Arts and
Ecology Center, permaculture program, and
WATER Institute director

Climate

Climate

A NEW STORY

CHARLES EISENSTEIN

North Atlantic Books
Berkeley, California

Published by
North Atlantic Books
Berkeley, California

Cover Photo © gettyimages.com/ivan101
Cover design by Jasmine Hromjak
Book design by Happenstance Type-O-Rama

Printed in Canada

Climate: A New Story is sponsored and published by the Society for the Study of Native Arts and Sciences (dba North Atlantic Books), an educational nonprofit based in Berkeley, California, that collaborates with partners to develop cross-cultural perspectives, nurture holistic views of art, science, the humanities, and healing, and seed personal and global transformation by publishing work on the relationship of body, spirit, and nature.

North Atlantic Books' publications are available through most bookstores. For further information, visit our website at www.northatlanticbooks.com or call 800-733-3000.

Library of Congress Cataloging-in-Publication Data

Names: Eisenstein, Charles, 1967- author.
Title: Climate : a new story / Charles Eisenstein.
Description: Berkeley, California : North Atlantic Books, 2018. | Includes
  bibliographical references.
Identifiers: LCCN 2018013922 (print) | LCCN 2018031946 (ebook) | ISBN
  9781623172497 (ebook) | ISBN 9781623172480 (paperback)
Subjects: LCSH: Environmentalism. | Environmental degradation—Prevention. |
  Climatic changes. | Global environmental change. | Nature—Psychological
  aspects. | Environmental psychology. | BISAC: NATURE / Environmental
  Conservation & Protection. | BODY, MIND & SPIRIT / Inspiration & Personal
  Growth. | POLITICAL SCIENCE / Public Policy / Environmental Policy.
Classification: LCC GE195 (ebook) | LCC GE195 .E39 2018 (print) | DDC
  363.7—dc23
LC record available at https://lccn.loc.gov/2018013922

1 2 3 4 5 6 7 8 9 MARQUIS 23 22 21 20 19 18

Printed on 100% recycled paper

North Atlantic Books is committed to the protection of our environment. We partner with FSC-certified printers using soy-based inks and print on recycled paper whenever possible.

Dedicated to the humble people, whose quiet service holds the world together.

Contents

Acknowledgments

This book is possible only because of the friends and allies who hold the field I write from and remind me I am not crazy. Among them are Bayo Akomolafe, Ben Phelan, Brad Blanton, Camila Moreno, David Abram, Frank Phoenix, Helena Norberg-Hodge, Gigi Coyle, Ian MacKenzie, Jodie Evans, Joshua Ramey, Kelly Brogan, Laurie Young, Lissa Rankin, Lynn Murphy, Manish Jain, Marie Goodwin, Matthew Monihan, Michael Lerner, Miki Kashtan, Orland Bishop, Pat McCabe, Polly Higgins, Satish Kumar, and so many more, some of them very dear. I also would like to give thanks to the near-strangers who shower me with generosity and encouragement; to the patrons who supported me financially during the years writing this book; and most especially to my wife, Stella, for her loyalty to my best self; to my parents, for fifty years of love; to my children, for showing me the future; and to my first wife, Patsy, for showing me the power of life to heal.

Prologue:

Lost in a Maze

Once upon a time, a man was lost in a maze. How he got there is another story—to learn a mystery, perhaps, or to find a treasure. In any case, by now he has forgotten how or why he came to be there. He holds onto a faint memory of a sunlit realm, or a memory of a memory, that tells him that the maze isn't the whole of reality. He got there somehow, and there must be a way out. And lately it has become more and more painful to be inside. The maze is getting hotter and hotter, and he knows he will die if he doesn't find the exit soon. What was once an exciting exploration has become a monstrous trap.

Frantically he races around seeking the way out. Left, right, left, right, up and down, around in circles he runs, hitting dead ends and turning back, finding himself again and again back at his starting point. He begins to despair—after all that effort he has gotten nowhere.

A committee of voices in his head offers him advice: how to run faster, how to choose smarter. He heeds first one, then another, yet no matter how different the advice the result is always the same. Sometimes, amid the cacophony, he hears another voice as well, a quieter voice telling him, "Stop." "You aren't getting anywhere," it says. "Just stop."

The other voices respond with outrage, "You cannot stop. You cannot rest. Only by using your two feet will you ever get out of here, and the situation is urgent so you'd better move those feet fast. The window of opportunity is closing. Now is the time for action. After you get out, then it will be time to rest."

And so he runs all the faster, his head filled with new stratagems, pushing himself to an all-out effort. And once again, after many twists and turns, he finds himself back in the center of the maze.

This time he has to stop. Out of sheer exhaustion and despair, he collapses in a heap. The tumult of advice subsides, leaving his mind for once quiet, as happens when every option is exhausted and one just doesn't know what to do. Now he has a chance to ponder his wanderings, and in the empty space of his mental quiet, new realizations are born. He realizes that there has been a pattern to his wanderings. Perhaps he followed each left turn with a right turn. He also remembers running past small, dark passages that he ignored because they seemed unpromising. He remembers glimpsing secret doors that he was too much in a hurry to investigate. In quietude, he begins to understand the structure of the territory he has been racing in.

By now his heartbeat and breath have calmed along with his mind, and another sound comes into his awareness. It is a beautiful, musical sound that, he now realizes, has been there all along, drowned out by his pounding footsteps and ragged breathing. He knows that he must never lose touch with that sound again.

The man begins to walk again, slowly this time. He knows that if he panics and starts running (understandable, since the crisis he's been trying to escape is real) then he will fall into old habits. Guided by his newfound understanding, he explores the small, dark passageways that he'd dismissed before. He enters the hidden doorways that take time to unlock. Sometimes these new doors and passages lead to dead ends too, but at least now there is hope. He is in new territory, unfamiliar territory. No longer is he endlessly finding himself back at his starting point. Now he is wandering for real.

As he leaves familiar territory farther behind, his understanding of the structure of that part of the maze becomes less and less useful. He faces choice points without a mental map. Should he turn left or right? At those moments he again stops, gets quiet, and listens, tuning in again to that musical sound that he keeps in his awareness. From which direction does the sound come through the most clearly? That is the direction he chooses.

Sometimes when he follows the music it seems to take him the wrong way. "That couldn't possibly be the way out," he thinks. But then the passage turns again, and he comes to trust more and more this sound that calls him.

Following the music, eventually the man reaches the final passageway, which he recognizes because of the daylight glow at the end of it. He emerges into the sunlit realm he always knew must exist; it is more beautiful than he ever dared imagine. And there he finds the source of the music. It is his Lover, who has been singing to him all this time.

1

A Crisis of Being

A Lost Truth

I still remember the event that made me into an environmentalist. I was seven or eight years old, standing outside with my father watching a large flock of starlings fly past. "That's a big flock of birds," I said.

My father told me then about the passenger pigeon, whose flocks once filled the skies, so vast that they stretched from horizon to horizon for hours on end. "They are extinct now," he told me. "People would just point their guns to the sky and shoot randomly, and the pigeons would fall. Now there aren't any left." I'd known about the dinosaurs before then, but that was the first time I really understood the meaning of the word "extinct."

I cried in my bed that night, and many nights thereafter. That was when I still knew how to cry—a capacity that, once extinguished through the brutality of teenage boyhood in the 1980s, was nearly as hard to resuscitate as it would be to bring the passenger pigeon back to earth.[1]

---

[1] The reader will notice that I sometimes capitalize "Earth" and sometimes do not. When I refer to it as a planet, I capitalize it. When I refer to it as a realm of habitation, or as a synonym for ground, or in the sense of a place where life happens (any planet like ours would be an earth), I do not.

These two kinds of extinction are related. From what state of being do we extinguish other species, ruin earth and sea, and treat nature as a collection of resources to be allocated for maximum short-term benefit? It can come only from the constriction, numbing, and diversion of our capacity to feel empathy and love. No mere personal failing, this numbing is inseparable from the deep narratives that run our civilization, and the social systems that those narratives support.

Appearances to the contrary, it is neither folly nor myopia that sets us on a path of collective ruin. These are symptoms of a deeper malady. Would you say of the alcoholic that if he were only shown that drinking harms his health, relationships, and economic security, then his dismal future would scare him into quitting? Of course not. The foolish sacrifice of the future for a temporary surcease from the inner pain isn't driven by stupidity. Therefore, you can harangue him about the damage to his liver all you want, and maybe he'll say, "Yeah, you're right," and cut back for a few weeks, or he will promise to drink less, with every good intention. But nothing will really change. How similar that scenario is to the climate talks. We agree to cut back—and agree, at the same time, to ignore the social and economic conditions that make cutting back impossible. Carbon emissions continue to grow after nearly three decades of climate talks and agreements. This pattern extends beyond the matters of climate. Species continue to perish, bat colonies and bee hives to collapse, forests to wither, coral reefs to bleach, and elephants and whales to die. No one wants to live on a barren planet, a sick planet, or a dying planet, yet like an addict we seem helpless to change course.

Like many clichés, "our addiction to fossil fuels" contains a lost truth. Usually I hear the phrase used in tones of condemnation or disgust (betraying the same lack of empathy that is part of the problem). But if we take the addiction metaphor seriously, we would next inquire as to what drives the addiction.

Some on the left say it is capitalism. Yet the Soviet Union committed grievous environmental damage as well; besides, capitalism (like communism) is itself embedded in more fundamental belief systems that are largely beneath the surface of our awareness. It is these that I intend to excavate in this book, hoping therefrom to derive precepts and strategies for ecological healing. I will describe how many of the efforts to fight climate change or save the environment are based on the same assumptions that drive us toward ruin. I will identify fundamental problems in what I'll call the Standard Narrative of climate change, and show how the framing of the problem is part of the problem. I will explain how solutions that come from that narrative risk making things worse. The maze thus revealed, I'll explore the dark passageways and secret doorways that the dominant discourse ignores, but that an alternative Story of the World illuminates.

It is not wrong ideas that drive addiction. Addiction arises in the presence of basic unmet needs. The food addict isn't really hungry for food; she is hungry for connection. The alcoholic is seeking just to feel okay for a while. The gambler yearns for liberation from economic or psychological confinement. The porn addict's true desire is for intimacy and acceptance. These (admittedly, trivialized) examples at least convey a general principle: Desire comes from unmet needs. When the true object of the desire is unavailable, the desire is displaced onto the most accessible substitute. What is the unmet need behind the addiction to fossil fuels?

In addiction theory there is a concept of addiction transfer: when the addict is forcibly deprived of the object of her addiction, she will transfer the addiction onto something else. Recipients of bariatric surgery who can no longer overeat might start drinking or gambling instead. Overeating, drinking, and gambling are symptoms of a deeper wound. Similarly, I will argue, the current environmentalist obsession with fossil fuels is also too narrow. Conceivably, we could find another fuel source and maintain the addiction to a system of economics and production that consumes the world.

What is it that we are really looking for in our quest for bigger, faster, and more? Later chapters on energy and agriculture make it clear that humanity's problems do not stem from any quantitative lack—hunger for instance is nearly always a result of maldistribution. We seek through growth to meet other needs, needs that, because they are fundamentally qualitative, growth can never meet. Basic human desires for connection, community, beauty, sacredness, and intimacy are met with faux substitutes that temporarily numb but ultimately heighten the longing. The trauma of our deprivation drives our collective addictions. Ecological healing therefore requires our society to look beneath its consumptive symptoms and reorient toward qualitative development. To do so requires significant reprogramming, since our guiding narratives, from economic to scientific, embody quantitative thinking.

Ecological deterioration is but one aspect of an initiation ordeal propelling civilization into a new story, a next mythology. By a mythology, I mean the narratives from which we weave our understanding of who we are, what is real, what is possible, why we are here, how change happens, what is important, how to live life, how the world came to be what it is, and what ought to come next. Ecological degradation is an inevitable consequence of the mythology—I call it the Story of Separation—that has dominated the last several centuries (and to an extent the last several millennia). To paraphrase Einstein, it will not be averted from within that mythology.

The essence of the Story of Separation is the separate self in a world of other. Since I am separate from you, your well-being need not affect mine. In fact, cast into an objective external universe, more for you is less for me; naturally then we are in competition with each other. If I can win the competition and dominate you, I'll be better off and you worse. The same goes for humanity generally vis-à-vis nature. The more control we can exercise over the impersonal forces of nature, the better off we will be. The more intelligence we can impose upon a random, purposeless universe, the better the world will be. Our destiny, then, is to ascend beyond nature's original limits, to become its lords and masters. The universe, this story says, is but atoms and void, possessing none of the qualities of a self that we experience as human beings: intelligence, purpose, sentience, agency, and consciousness. It is up to us then, to bring these qualities to the dead building blocks of the universe, its generic particles and impersonal forces; to imprint human intelligence onto the inanimate world.

The Story of Separation reverberates through every institution of the modern world. In other books I've described how it underlies money, law, medicine, science, technology, education, etc., and how these institutions might evolve under a different story.

This book also aims to describe and, I hope, accelerate the transition to a new (and in many ways ancient) story, with specific reference to climate change and the environmental crisis generally. A shift in mythology is more than a cognitive shift. In this book I will argue that the external changes we face are far more profound than merely switching industrial society to a zero-carbon fuel stock. Every aspect of society, the economy, and the political system must come into alignment with a new story.[2]

The name I like to use for the new story is Thich Nhat Hanh's term "interbeing." Although the word has Buddhist overtones, I do not profess to be a Buddhist, nor need the reader embrace Buddhism to appreciate the insights the concept allows.

Interbeing doesn't go so far as to say, "We're all one," but it does release the rigid boundaries of the discrete, separate self to say that existence is relational. Who I am depends on who you are. The world is part of me, just as I am part of it. What happens to the world is in some way happening to me. The state of the cultural climate or political climate affects the condition of the geo-climate. When one thing changes, everything else must change too. The qualities of a self (sentience, agency, purpose, an experience of being) are not confined to humans alone. And the results of our actions will come back to affect ourselves, inescapably.

Interbeing must be more than a philosophical concept if anything is going to change. It must be a way of seeing, a way of being, a strategic principle, and most of all a felt reality. Philosophical arguments alone will not establish it any more than appeals to prudence and reason will solve the ecological crisis.

When we restore the internal ecosystem, the fullness of our capacity to feel and to love, only then will there be hope of restoring the outer. Each

---

2   I use the adjective "new" to mean "new for modern civilization as a guiding narrative." In fact it is not at all new. Not only did older, indigenous cultures hold some version of the Story of Interbeing, it inhabits Western civilization as well in the form of esoteric teachings, wisdom traditions, and cultural countercurrents. What would be new would be a mass civilization operating according to the principles of interbeing.

level of healing proceeds apace, just as each form of extinction mirrors the rest. That is not to suggest we withdraw from outer activism in favor of inner cultivation. It is that love and empathy are the felt dimensions of the Story of Interbeing, and we cannot act effectively from that story, nor truly serve it, without their guidance. They are the song that will lead us out of the maze. To follow their guidance we must regain our listening capacity, which trauma and ideology have numbed and restricted to a very narrow bandwidth.

Then we will know how to change the systems that reify Separation by severing our ties to community, plants, animals, land, and life and replacing those ties with the technology-mediated, money-mediated, generic relationships of mass society. (Thus bereft, no wonder we always hunger for "more.")

Love is the expansion of self to include another. In love, your well-being is inseparable from my own. Your pain grieves me and your happiness gives me joy. The ideology of modernity circumscribes the scope of our love by assigning a narrow identity to the self and relegating the nonself to the status of mute, insensate objects or self-interested competitors. To care about others beyond their utility to oneself becomes therefore something of a delusion, like loving your pet brick.[3] Perhaps that is why so much environmental rhetoric comes in the form of warnings that bad things will happen to us if we don't change our ways. We call arguments "rational"

---

[3] Full disclosure: I have a special brick that I use for my *qigong* practice toward which, I must confess, I feel quite some affection. As for the irrationality of love, I'll quote a bit of doggerel from the science fiction writer Isaac Asimov. Do you detect as I do a note of abject defeat underneath its know-it-all cheekiness?
Tell me why the stars do shine,
Tell me why the ivy twines,
Tell me what makes skies so blue,
And I'll tell you why I love you.

Nuclear fusion makes stars to shine,
Tropisms make the ivy twine,
Raleigh scattering make skies so blue,
Testicular hormones are why I love you.

when they appeal to self-interest. This book will argue that rational reasons are not enough; that the ecological crisis is asking for a revolution of love.

For the discrete and separate self in a world of other, love is irrational. Steeped in the logic of separation, the mind is ever in conflict with the heart. Not so, in the logic of interbeing, which recognizes that what happens to the other, to the incarcerated, to the bombed, to the trafficked, to the clear-cut, to the polluted, and to the extinguished is happening, in some sense, to the self as well. In the Story of Interbeing, heart and mind are reunited, and love is what the truth feels like.

If love is truth, then the source of our apparent myopia is clear. It is love benumbed. We do not see that what we devalue and destroy is part of ourselves. We do not see that we aren't merely conditionally dependent on the oceans, rainforests, and every living system on Earth for survival; that something more important than survival is at stake. It is our humanity. It is our full beingness. Love benumbed, we believe that we can inflict damage without suffering damage ourselves.

Of course, I would not write a book that were just a vague promise that love will save the world. How do we enact it systemically? How do we overcome what blocks it? How do we awaken our benumbed empathy? How do we translate the diagnosis I've offered into practical action on the level of politics and ecological healing? These questions are the subject of this book.

The Identity of "They"

Species extinction, as you know, did not end with the nineteenth century. The fate of the passenger pigeon foreshadowed the calamity that is now overtaking life on this planet, a calamity that has left none of us untouched. The calamity is the impoverishment of life, in every sense of that phrase. Extinction is one kind of impoverishment; the more general decline in biodiversity is another; so also with the spreading deserts on land and in the ocean and the general depletion of life even where it is green. Even when species don't go extinct, often they decline to small remnant populations, shrink to a small portion of their original range, lose subspecies and genetic diversity, and inhabit vastly simplified ecosystems. This withering of biological life accompanies the impoverishment of human life and cultural vitality. All partake of the same crisis.

I recently made the acquaintance of a farmer in North Carolina whom I'll call Mike, a man of the earth whose family has been here for three hundred years. His thick accent, increasingly rare in this age of mass-media-induced linguistic homogenization, suggested conservative "Southern values." He was indeed full of bitterness, though not against the usual racial or liberal suspects; instead he launched into a tirade about the guvmint, chemtrails, the banks, the 9/11 conspiracy, the apathy of the "sheeple," and so on. "We the people have got to rise up and smash them," he said, but there was no fervor in his voice, only a leaden despair.

Tentatively, I broached the idea that the perpetrators of these crimes are themselves imprisoned in a world-story in which everything they do is necessary, right, and justified; and that we join them there when we adopt the paradigm of conquering evil through superior force. That is precisely what motivates the technologies of control, whether social, medical, material, or political, wielded by those we would overthrow. Besides, I said, if it comes down to a war to overthrow the tyrants, if it comes down to a contest of force, then we are doomed. They are the masters of war. They have the weapons: the guns, the bombs, the money, the surveillance state, the media, and the political machinery. If there is hope, there must be another way.

Perhaps this is why so many seasoned activists succumb to despair after decades of struggle. Dear reader, do you think we can beat the military-industrial-financial-agricultural-pharmaceutical-NGO-educational-political complex at its own game?[4] The modern environmental movement, and especially the climate change movement, has attempted just that, not only risking defeat but sometimes worsening the situation even in its victories. The ecological crisis is calling us to a deeper kind of revolution. Its strategy involves restoring what the modern worldview and its institutions

---

[4] I hope I haven't left anyone out—I don't wish to be rude.

have rendered nearly extinct: our felt understanding of the living intelligence and interconnectedness of all things. To not feel that, is to be not fully alive. It is to live in poverty.

Mike wasn't understanding me. He is an intelligent man, but it was as if something had possessed him; no matter what I said, he would pick up on one or two cue words to pour forth more bitterness. Obviously, I wasn't going to "defeat the enemy" by force of intellect (thus enacting the very same paradigm I was critiquing). When I saw what was happening, I stopped talking and listened. I listened not so much on a semantic level, but to the voice beneath the words and to all that voice carried. Finally I knew what to do. I asked him the same question I want to ask you: "What made you into an environmentalist?"

That is when the anger and bitterness gave way to grief. Mike told me about the ponds and streams and wild lands that he hunted and fished and swam and roamed in his childhood, and how every single one of them had been destroyed by development: cordoned off, no-trespassed, filled in, cut down, paved over, and built up.

In other words, he became an environmentalist in the same way that I did, and, I am willing to guess, the same way you did. He became an environmentalist through experiences of beauty and loss.

"Would the guys ordering the chemtrails do it, if they could feel what you are feeling now?" I asked.

"No. They wouldn't be able to do it."

The truth of that moment Mike and I shared stands alongside the reality that, actually, they *would* be able to do it, that "they" in fact includes each one of us who participates in this civilization. A single moment of reverence, gratitude, or grief, however profound, is not enough to undo generations of programming, nor to extricate ourselves from an economy and society of ecocide. Are you able to get into your car, knowing the effect of emissions and oil spills and the geopolitics of oil extraction? I certainly am, and you probably are too. You might have a story about why it is okay, why in your case it is justified, or at least why *you* are okay for doing it. "I have no choice," you might think. Or "At least I feel bad about it. At least I'm opposed to it. At least I vote for people or donate money to organizations who are trying to change the system. Besides, I'm driving a hybrid." All kinds of reasons why it is okay to get into your car right now. Or maybe you don't think about it at all.

My point here is not that you are deluding yourself—you pathetic, self-justifying hypocrite! It is to illuminate the fallacy of our judgments and the war thinking that they engender. And it is to suggest that we are not normally feeling what Mike was describing, because we live in a system, an ideology, and probably a wounded psychology that allow full feeling only sporadically. The system numbs us; it also depends on our numbness.

I want us to transcend the language of "Is it okay?" entirely, and underneath it, "Am I okay?" This is the language of war turned inward. Along with defeating the enemy, we seek to conquer its internal projection: the greedy, hypocritical, dishonest, egotistical, self-serving parts of ourselves. In this campaign, self-disgust is considered an ally, the first sign of redemption, because now we are joining the good side, with parts of ourselves as the enemy. Dissociating from those parts, we imagine that we are making progress in overcoming them. Such great efforts we are making, such commendable progress.

Are we ever making progress though? Or is any progress merely in our ability to excuse, cloak, and rationalize the choices that don't fit the image of our ethics?

Corporations and governments do just that: they cloak, they excuse, they deny, and they make cosmetic, self-justifying changes to uphold a green image. We would like to blame greenwashing on corporate duplicity and greed—giving us an external enemy to fight—but (like our own self-justifications) I am afraid it is rooted in something much deeper.

In both cases, personal and political, to blame moral failings for the horrifying predicament of people and planet is a dangerous error that diverts attention away from systemic and ideological causes. It disguises a problem that we don't know how to solve as a problem that we do. We know, in theory at least, how to stop bad people from doing bad things. We can deter them, surveil them, imprison them, or kill them. We can fight them, and if we win the fight, the problem is solved.

Our political discourse is rife with good-versus-evil narratives. It is obvious to each side that they are good and the other side is evil (or some cipher therefor: sick, irrational, twisted, unethical, corrupt, "acting from the reptilian brain," etc.). Both sides agree on that. Therefore, both sides also agree on the strategic template for victory: arouse as much outrage and indignation as possible among the Good Folk so that they will rise up and cast down the Evil Folk. No wonder our civic discourse has degenerated into such polarized extremes.

That does not mean that I hold no opinion about which side is right in the political questions of our day. Nor am I saying that truth is a matter of opinion or that we create our reality. It is rather that those in our society typically misunderstand the causes of others' opinions and behavior.

To blame evil is to misdiagnose the problem. I explored this idea in depth in my last book; here I'll just ask you to insert yourself into the totality of the circumstances of a fracking executive. The "totality of circumstances" could include:

- The corporate culture
- The culture of the energy industry
- Performance pressure
- Economic pressures on the business, rooted in the economic system
- Years of attacks from hostile "enviros" who appear to you to be ignorant and misguided
- Stories of "American energy independence"
- Ideologies of progress, growth, and technology
- The perceptual set of Earth-as-thing
- Childhood "success" programming

What actions would you take from those conditions? What would be your hardest choices? Your most painful compromises?

What are your hardest choices and most painful compromises right now? Do you drive a car that burns gasoline? Did you drive somewhere when it was raining yesterday, when you really could have biked? Do you take energy-consuming hot showers? Do you tread on cement sidewalks? Do you use a cellphone containing conflict minerals? Do you use credit cards or banks who fund the pillage of nature? If so, someone out there probably thinks you are evil too. Exploiter! Hypocrite! Consumer of more than your share! You might sometimes think that about yourself too. Other times, you will have compassion for yourself, realizing that given your circumstances, your burdens, your traumas, and your limitations, you are doing the best you can.

Does this mean we might as well give up on change? No. It means we need to ask, What are the circumstances that give birth to the choices that are harming the world? Engaging other people, we have to ask the question that defines compassion: What is it like to be you? The more we understand, the more we live in reality and the less we inhabit a fantasy world populated by our projections. You can go ahead and see your opponents as dastardly villains, but if that is not the truth of who they are, then you are living in a delusion. Focusing on the bad guys, we become blind to deeper, systemic causes, forever chasing false solutions that actually maintain the status quo.

Living in a delusion, we endlessly re-create its landscape; we repeatedly enact its roles and manufacture its dramas, racing along the same old paths of the maze. Even if we achieve temporary victory against the bad guys, the overall situation doesn't seem to change. We never get closer to the exit. What we normally achieve is, instead of victory, a strengthened conviction that we are in fact the good guys. That polarized view is one of the things we will have to give up if we are to launch the era of ecological healing. Are you prepared to sacrifice being the winner? Are you willing to sacrifice being one day proved right? Are you willing to stop seeing yourself on Team Good fighting Team Evil? Because that's one thing that both sides of any debate normally believe of themselves, and that is the template of "othering" that exemplifies and reinforces human separation from nature.

I ask these questions deliberately. I will argue in this book that all the positions on the spectrum of climate change opinion, from skepticism to catastrophism, are wrong. Like those who blame evil people for the world's evil, they operate in too shallow a causal framework. The totality of circumstances driving ecological degradation and climate derangement is greater than conventional opinion recognizes.

The Fight

None of the above is to deny that horrible things are happening to life on this planet. Someone is bulldozing the trees, draining the wetlands, bottom-trawling the fish, and polluting water, air, and soil. Each time, that someone is a human being.

Since most of the damage happens at the behest of large corporations, it seems reasonable to name them as the enemy. Expose their immoral behavior! Hold them to account! Deter their crimes with meaningful penalties! Get their money out of politics! Then we can at least reduce their worst excesses.

This argument is reasonable under current conditions, but it accepts as unchangeable the very things we must change. I'll offer some specifics later in the book; for now, a generality: fighting the enemy is futile when you inhabit a system that has the endless generation of enemies built into it. That is a recipe for endless war.

If that is to change, then one of the addictions—more fundamental than the addiction to fossil fuels—that we are going to have to give up is the addiction to fighting. Then we can examine the ground conditions that produce an endless supply of enemies to fight.

The addiction to fighting draws from a perception of the world as composed of enemies: indifferent forces of nature tending toward entropy, and hostile competitors seeking to further their reproductive or economic self-interest over our own. In a world of competitors, well-being comes through domination. In a world of random natural forces, well-being comes through control. War is the mentality of control in its most extreme form. Kill the enemy—the weeds, the pests, the terrorists, the germs—and the problem is solved once and for all.

Except that it never is. World War I—the "war to end all wars"—was followed by another, even more horrific, soon after. Nor did evil disappear after the defeat of the Nazis or the fall of the Berlin Wall. The collapse of the Soviet Union was, however, a crisis for a society that had come to define itself through its enemies; thus followed a desperate search for a new enemy in the early 1990s that resulted in the feeble candidate of "Colombian drug lords" for a time before settling on "terror."

The War on Terror gave a new lease on life to a culture built on warmaking; indeed it seemed to offer the prospect of permanent war. Unfortunately for the military-industrial complex, the public seems to be growing less terrified of terror, necessitating a series of new threats by which to maintain a climate of fear. It is hard to say that the scare campaigns of the last few years—Russian hackers, Islamic terror, Ebola, the Zika virus, Assad's chemical weapons, and Iran's nuclear program, to name a few—have not worked. The media at least clangs the alarm, and the public seems to have gone along with the policies that these campaigns justify, such as massive spraying programs in Florida to "combat Zika." However (and this may in part be a function of my countercultural social circles), I have not seen much actual fear about these things, nothing like the palpable dread of the Soviet Union that was nearly universal in my childhood. The public discounts pretty much anything the authorities say, including the fear-mongering. Its apathy permits the governing elites to pursue their programs of control, but no longer do they channel and harness real fear. Is anyone outside the political classes actually afraid of Iran, Bashar al-Assad, or Vladimir Putin? One suspects that neither are the politicians, though they may display a semblance of alarm as a political posture.

I bring up the waning power of scare tactics because the effort to halt ecological collapse uses many of these very same scare tactics. The primary climate change narrative is basically, "Trust us, bad things will happen if we don't hurry up and make big changes. It's almost too late—the enemy is at the gates!" I want to question the assumption that we can and should motivate the public with fear-based appeals to self-interest. What about the opposite? What about appeals to love? Is life on earth valuable or sacred in its own right, or only in its utility to ourselves?

Climate change activism abounds in war narratives, war metaphors, and war strategies. The reason, aside from the deep-seated habits of the Story of Separation, is the desire to inspire the fervor and commitment that people display in wartime. Following the rhetorical template of war, we invoke an existential threat.

I don't think it is working. I hesitate to use the term "climate change" in my essay titles. The last time I did so, one reader wrote to me saying, "I almost didn't read your post because it had the words climate change in the title, and I'm just so sick of hearing the same thing over and over again."

Maybe we are becoming war-weary. Does it take more and more exhortation to goad you into joining another battle? Have you encountered burnout, when no new horror can stimulate you to the kind of engagement you practiced a few years ago? Burnout seems the downfall of activists, but as the story of the man lost in a maze implies, it can be a necessary initiation into a wholly different mode of engagement.

My friend Pat McCabe, a Diné (Navajo) woman and longtime student of the Lakota Way, puts it this way: "When you reach the end of your resource, then the magic happens." When we exhaust what we know, then what we don't know becomes possible.

Struck with grief at the ruin of life on earth, one might understandably take offense at any suggestion that we "give up the fight." To someone steeped in war mentality, to give up the fight means to withdraw from action. I'm suggesting we give up the fight in another sense: as the orienting principle of our efforts to heal the earth. There may still be battles, but we will access much greater power to heal when we frame the issue in terms of peace.

It is often observed that the last major war to unambiguously achieve its objectives was World War II. Since then, military conflicts have usually ended in stalemate, quagmire, or defeat for the stronger power. The failure of, for example, the U.S. war in Afghanistan is not due to inferior weaponry. It is that its weaponry is insufficient to its objective, which cannot be achieved by force. Guns and bombs cannot usually bring stability, "win hearts and minds," or make a country pro-American, unless it is an unambiguous case of saving people from evil despots or aggressors.[5] To justify war, we have to fit every situation into that storyline, as the media has tried to do in every conflict since Vietnam.

The same goes for nonmilitary wars. In my lifetime I've heard declarations of a War on Poverty, a War on Cancer, a War on Drugs, a War on Terror, a War on Hunger, and now a War on Climate Change. None of these have been any more efficacious than the War in Iraq.

If the "fight" against climate change is a war, it is clear which side is winning. Greenhouse gas emissions have relentlessly increased since they were first widely acknowledged as a problem in the late 1980s. Deforestation has also continued and in some places even accelerated since then. Nor has any progress been made in altering the basic fossil-fuel-dependent infrastructure of society. If war were the only answer, then we would have to respond by fighting even harder. If there is another way, then the habit of fighting becomes an obstacle to victory.

In the case of ecocide, the mentality of war is not only an obstacle to healing, it is an intimate part of the problem. War is based on a kind of reductionism: it reduces complex interconnected causes—that include oneself—to a simple, external cause called the enemy. Furthermore, it normally depends on the reduction of the enemy to a degraded caricature of a human being. The demonization and dehumanization of the enemy is little different from the desacralization of nature upon which ecocide depends. To render nature into an *other* undeserving of reverence and respect, an object to dominate, control, and subjugate, is of a kind with the dehumanization and exploitation of human beings.

---

[5] Some argue that the true objective of recent wars has been to sow chaos and destroy the ability of sovereign governments to resist neoliberal free trade policies and imperialist geopolitical objectives. In that analysis, some wars such as the one that dismembered Yugoslavia or destroyed Libya were a great success. Nonetheless, the point still applies that the tools of war are becoming impotent to achieve what we say and believe we want.

Respect for nature is inseparable from respect for all beings, including the human. It is impossible to cultivate one without the other. Climate change, therefore, calls us to a greater transformation than a mere change in our energy sources. It calls us to transform the fundamental relationship between self and other, including but not limited to the relation between the collective self of humanity and its "other," nature.

The philosophically inclined reader may protest that self and other are not really separate, or that the human/nature distinction is an artificial, false, and destructive binary, an invention of the modern mind. Indeed, "nature" as a separate category suggests that we humans are unnatural and, therefore, potentially exempt from nature's laws. Whatever the underlying metaphysics, what is changing is our *mythology*. We never were separate from nature and never will be, but the dominant culture on earth has long imagined itself to be apart from nature and destined one day to transcend it. We have lived in a mythology of separation.

Part of the mythology of separation is a belief in nature-as-thing; in other words, the belief that only human beings are possessed of full selfhood. This is what licenses us to exploit the beings of nature for our own ends, much as dehumanization of brown people licensed lighter-skinned people to enslave them.

The dominant culture's recognition of who counts as a fully subjective, conscious, and worthy self has been expanding now for several hundred years. Two or three centuries ago, only a propertied white male was a full subject. Then that category was expanded to include all white males. Eventually it expanded again to include women as well, and people whose skin is not white. Then along came the animal rights movement, which said that animals too have consciousness, subjectivity, and an inner life, and should not therefore be treated as mere brutes or meat-machines. More recently, remarkable scientific discoveries have emerged around plant intelligence, mycelial intelligence, soil intelligence, forest intelligence, and even the capacity of water to hold and transmit complex, dynamic patterns of information. These discoveries seem to be converging on the universal indigenous belief that everything is alive and aware.

Just as bigotry and ecocide both depend on the dehumanization or "de-selfing" of the other, so also is the reversal of both part of the same movement toward a Story of Interbeing. Again, that term goes beyond mere interconnectedness or interdependency, to say that we are existentially connected to all other beings and to the world at large. My very being partakes in your being, and in the being of the whales, the elephants, the forests, and the oceans. What happens to them happens as well to me, on some level. When a species goes extinct something dies in us too; we cannot escape the impoverishment of the world we live in.

This applies equally to ecological, economic, and political well-being. The days of colonialism and imperialism—in which the wealth of one nation was built on the plunder of others—are dwindling. The era of thinking that human wealth could be built on the plunder of nature is nearly over too. Certainly, the outward structures of both kinds of plunder seem as robust as ever, even expanding to new extremes. However, their ideological core has hollowed out. Our converging crises are initiating humanity into the new and ancient mythology of interbeing.

Later I will argue that the reality of the climate crisis is different from our common perception of it. Yet perceptions are important. The core truth of climate change is that we are at the end of an era. We are at the end of the Age of Separation. It is a transition that has been under way for three generations now, inaugurated by the most extreme of all possible technologies of control applied at the very pinnacle of Total War. I am speaking, of course, of the Bomb.

The Age of War properly came to an end in 1945, when for the first time in history human beings developed a weapon too terrible to use. It took two horrific applications of the atomic bomb to set the stage for decades of "mutually assured destruction," a glimmering of the evolutive realization that what we do to the Other, we do to ourselves. For the first time in history, total war between the great powers was impossible. Today, aside from an unregenerate minority, no one contemplates using nuclear weapons even in cases where retaliation is unlikely. Radioactive blowback makes large-scale use unthinkable, but there is another thing that holds us back too. We name it, perhaps, conscience or ethics, but history makes it tragically clear that conscience or ethics alone are not enough to stop the foolish and the horrific. No, something else has changed.

What has changed, I believe, is that the consciousness of interbeing is dawning in the dominant civilization. What we do to the Other, we do to ourselves. This will be the defining understanding of the next civilization—if there is a next civilization. Right now we (usually in this book, when I say "we" I mean the dominant culture on this planet) are facing lesson number two in the curriculum of interbeing. Lesson one was the Bomb. Lesson two is climate change.

2

Beyond Climate Fundamentalism

Does Nothing Else Matter?

"Someday, Charles, you are going to have to decide if you want to be relevant."

So said to me a prominent environmentalist after hearing me describe the diverse fields of my activity and interest. What he meant was something like this:

> There is a shrinking window for climate action before irreversible feedback loops render human extinction inevitable. Therefore, the only relevant action you can take right now is to put 100 percent of your efforts into cutting greenhouse gas emissions as swiftly as possible by whatever means necessary. Your other interests are irrelevant. If we don't implement a meaningful carbon tax soon, then the healing of the relation between the masculine and feminine won't matter. Nor will saving the whales. Nor will ending the school-to-prison pipeline. Social justice, education, psych meds, holistic medicine, scientific anomalies, attachment parenting, community building, new economics, philosophy, history, cosmology, neo-Lamarckian biology, sacred plant medicines, nonviolent communication, plant intelligence, threatened languages, indigenous sovereignty, pansubjective metaphysics—none of the issues you write about matter unless they have a direct, significant, near-term impact on greenhouse gases. Once we've won that fight, we can turn our attention to those other things. So are you going to join the fight?

This pattern of thinking is called fundamentalism, and it closely parallels the dynamics of two defining institutions of our civilization: money and war. Fundamentalism reduces the complex to the simple and demands the sacrifice of the immediate, the human, or the personal in service to an overarching ulterior goal that trumps all. Disciplined by the promise of heavenly rewards or hellish punishments, the extreme religious fundamentalist shuts down his humanity in service to what his religion says God wants. Disciplined by economic exigency, millions of people sacrifice time, energy, family, and what they really care about in pursuit of money. Disciplined by an existential threat, a nation at war turns away from culture, leisure, civil liberties, and everything that is of no utility to the war effort.

Anyone who is wary of these institutions might also be wary of the standard climate change narrative, which lends itself to the same focus on a universal cause and the same mentality of sacrifice to an all-important end. If we agree that the survival of humanity is at stake, then any means is justified, and any other cause—say reforming the prisons, housing the homeless, caring for the autistic, rescuing abused animals, or visiting your grandmother—becomes an unjustifiable distraction from the only important thing. Taken to its extreme, it requires that we harden our hearts to the needs in front of our faces. There is no time to waste! Everything is at stake! It's do or die! How similar to the logic of war. No wonder, as a community organizer just told me, there is such hostility toward environmentalists among inner-city and other impoverished populations. They are the ones whose needs are ignored and indeed who are sacrificed first in the war effort.

While this book is focused on the realm of ecological healing, it disengages from the rhetoric of "Nothing else is important compared to this." That's the rhetoric that has alienated so many working-class people and minorities from environmentalism, because it carries a patronizing message of "We know better than you do what you should be caring about." It invalidates their grievances. Because, really, what does racially biased stop-and-frisk policing and the criminalization of large segments of the population matter in the face of civilizational collapse? What does sweatshop labor or carcinogens in the water supply matter, when climate change might render Earth inhospitable to human life? Your concerns are not important. If we carry this belief, even without being so impolitic as to voice it, we are going to radiate a crusading energy that is attractive only to our fellow fundamentalists.

If we want to foster a broad social consensus to protect and heal the planet, then we need to undo this logic at the source. The mind that is steeped in Separation protests, "But it is true! None of these things are relevant if the atmosphere warms by ten degrees."[1] This belief depends on a world-story that does not recognize the intimate interconnectedness of all things. If we see reality as a collection of separate, causally dissociated phenomena, then of course it will seem that stopping gentrification in Brooklyn or sex trafficking in Haiti is frivolous in the face of climate change.

From the Story of Interbeing, we intuit different kinds of cause and effect. We are not surprised that in a carceral society that locks up millions of its members, those outside the prisons lose their freedom too. We are not surprised that when a nation perpetrates violence around the world, that no amount of security, surveillance, walls, or fences can keep violence from sneaking back in, as domestic violence or self-destructive habits. And we are not surprised that environmental pollution and habitat degradation are mirrored in bodily illness and the degradation of our inner landscapes. The illusion of separation has us think that one could conceivably thrive on a poisonous planet with the right air filters, water filters, EMF blockers, supplements, air conditioners, antibiotics, antifungals, bug zappers, and so on, replacing a world of nature with a world of technology. In interbeing, we know that health for one is impossible to sustain without health for all.

---

[1] Degrees Celsius. In this book I will use the metric system unless otherwise noted. Actually I prefer traditional measures for everyday use, and the metric system for scientific applications. Traditional measures are less arbitrary; they relate to human experience (knuckles, feet, strides, a very cold or hot day, etc.). The metric system, in contrast, wipes out local and cultural differences and replaces them with a global standard. Like the commodification of nature and culture, this has been seen as progress.

If we want solidarity, we need to understand that genocide and eco-cide, human degradation and ecological degradation, are part of the same fabric, and that neither will change without the other changing. It is not that we should pay attention to racial or class injustice with the strategic goal of bringing those people into environmental activism. It is to recognize that healing on any level contributes to healing on every level. Because we are unaccustomed to thinking holistically, it seems counterintuitive that starting a social enterprise that employs homeless people will help stop climate change; the causal links are not evident to our way of seeing. Our dominant system of knowledge production (science) operates by controlling variables, breaking wholes into parts, and establishing measurable, predictable causal mechanisms. Then the knowledge is culturally legitimate. But the causal threads that link homelessness to ecological ruin are neither measurable nor predictable. Indeed, a cynic, channeling Ebenezer Scrooge, might argue that rehabilitating homeless people worsens climate change by transforming them into consuming members of society.

Of course, it is possible to construct an argument that housing the homeless contributes to the health of the biosphere, but it will not easily fit into the language of climate policy, nor is it likely to convince Mr. Scrooge. However, when Scrooge undergoes a shift in consciousness and sees the world through the eyes of interbeing, he will expect that the two phenomena are related. Believing in an innate intelligence pervading all phenomena, he might surmise that a society inhospitable to its vulnerable members will be mirrored by a planet that is inhospitable to society. He will expect that the deep roots of homelessness are common with the deep roots of climate change. Instead of "fighting homelessness" he will seek to understand the bed from which it arises. He will understand that it is okay to devote himself to what stirs his compassion the most, confident that what he is doing is still "relevant" in the face of global crisis. And he will no longer operate from self-preservation and survival anxiety, because he will understand that his well-being is inseparable from that of all in his expanding circle of love.

The question to explore then is what induces a shift to the consciousness of interbeing? Scrooge's creator Charles Dickens knew. It is through a confrontation with beauty, suffering, and mortality. It is through a connection to what is real. One might call it an initiatory experience. Without it, the grip of self-preservation and survival anxiety never loosens. We might try to leverage those fears (through the threat of climate change) to motivate pro-environmental behavior, but invoking self-interest to solve a problem caused by runaway, blind self-interest merely adds fuel to the fire. We need the opposite: to expand the circle of compassion to include every being on this earth.

The Perverse Consequences of Carbon Reductionism

Climate fundamentalism, translated into policy, bears consequences that are in direct opposition to what those policies seek to achieve. The main problem lies precisely in the aforementioned reductionism—to simplify a complicated matrix of causes into a single, identifiable cause. In today's environmental discourse, that cause is greenhouse gases, in particular carbon dioxide.

Like war thinking and money thinking, the problem with carbon reductionism is that it reduces "everything matters" to "one thing matters." In the words of Moreno and colleagues, "Once species and ecosystems have been entered into accounts, there is no need to look further into complexities, uncertainties and interlinkages ... trying to make reality and its contradictions fungible into carbon units entails cultural, symbolic and epistemic violence."[2]

Carbon reductionism sits comfortably within a broader, scientific reductionism. The indictment of science as reductionistic is often misunderstood to refer to its quest to explain the behavior of wholes by the properties of their parts. This quest, though, rests on a more insidious and more fundamental reductionism: that of the world into number. Its conceit is

---

[2] Moreno et al. (2015).

that someday, when everything has been ordered, classified, and measured, we will have penetrated every mystery and the world will finally be ours. This reduction of reality to quantity is a reduction of the infinite to the finite, the sacred to the mundane, and the qualitative to the quantitative. It is the abnegation of mystery, aspiring to encompass all of reality in its bounds.

The totalizing quest to capture the world in number never succeeds. Something always escapes the metrics and the models: the unmeasurable, the qualitative, and what seems irrelevant. Usually, the judgment as to what is relevant encodes the intellectual biases of those doing the measuring, and often the economic and political biases too. You might say that what is left out is our shadow. Like many things we ignore or suppress, it roars back in the form of perverse, unforeseeable consequences. Thus, although it is the epitome of rationality to make decisions by the numbers, the results often appear to be insane.

To see the problem, consider the Tehri Dam project on India's Bhagirathi River, completed in 2006. Constructed after decades of opposition by environmentalists and local residents, the dam submerged pristine ecosystems and ancient farms, displacing a hundred thousand villagers. Like countless other dams still being built in India, China, and Africa, it was touted for its contribution to greenhouse gas reduction and has been one of many dams to generate carbon trading credits. On a superficial level, it attained its measurable objective. But what about the displaced villagers? It could be that in the particulars that are measured, their lives improved: perhaps each was rehabilitated in concrete apartments superior to their ancestral homes in terms of square meters, plumbing, and electrification. However, in terms of the lost traditions, severed social ties, lost memories, lost knowledge, and the uniqueness of each submerged place—in short, in terms of all that could not be measured, and all that was considered not worth measuring—human beings and nature suffered a grievous loss.

Adding injury to injury, in the long run it is doubtful whether the dam even reduced CO_2 levels. Before they were displaced, the villagers had nearly a zero carbon footprint, or perhaps a negative footprint given that traditional agricultural practices can sequester carbon in the ground. Following their displacement, the newly urbanized villagers had to adopt more carbon-intensive consumer lifestyles, eating food shipped in from distant places, getting jobs in the industrial economy. Further, each new hydroelectric dam contributes to a trend of industrialization, adding to an infrastructure that is always hungry for more. It didn't come in place of coal-fired plants; it came in addition to them.

Hydroelectric dams generate electricity without burning fossil fuels, it is true, and it is easy to compute the tons of CO_2 that would be emitted by equivalent coal- or gas-fired plants. It is much harder to compute the carbon storage capacity of the ecosystems submerged in the dam reservoir, or the methane released by inundated vegetation (although recent estimates put methane emissions from artificial reservoirs at 104 megatons annually—as much as all fossil fuel methane emissions combined).[3] Harder still to calculate would be the effects of trophic cascades initiated by the deprivation of organic sediments to fish and riparian ecosystems downstream. The sediment is essential to build deltas and prevent ocean encroachment.[4] Given wetlands' huge carbon sequestration potential, it is possible that (even within the carbon reductionism frame, not to mention the water frame I'll present later) dam removal contributes more to climate stability than dam building does. Our "science-based" opinion depends on what we include in our measurements.

One unfortunate result of the fossil fuel divestment movement has been a giant land grab in Africa and South America, as investment capital turns toward biofuels plantations. Biofuels represent the most extreme possible form of reductionism: the reduction of living beings to heat. Along the way, existing peasant agriculture and ecosystems are also reduced—to jatropha or palm oil plantations, sugar cane plantations, woodchipping operations,

---

3   Magill (2014).

4   Robbins (2017).

and so on—even as diverse farm livelihoods are reduced to wage labor. By way of illustration, in the last decade controversy has erupted over the acquisition of vast tracts of land in Ghana by European corporations for the purpose of planting jatropha, whose oil-rich seeds, while toxic to humans and animals, are an excellent biofuel feedstock. Jatropha requires large plantations (1,000+ hectares) to be economically viable, which must be cleared of existing vegetation. Usually, they must be cleared as well of existing smallholder farmers. Since most land in Ghana is communally owned, this requires making deals with traditional chiefs, often illiterate, who may not understand the legal ramifications of the documents they are thumbprinting, especially when they are accustomed to regarding land as a sacred being rather than a fungible commodity.

The result is massive disruption of traditional lifestyles, human rights abuses, hunger, and ecological degradation. In a story replayed around the world, one reads of farmers showing up one day at their fields only to be told they are trespassing, and must abandon years or decades of investment in the land. The biofuel companies say that only previously uncultivated land is used and (somewhat contradictorily) that farmers who are evicted are compensated, but these claims don't always coincide with facts on the ground. Traditional chiefs or other persons of influence may get hired by the biofuel companies, setting their interests against those of the community. The plantation jobs dangled in front of the community don't always materialize, nor are they sufficient to compensate for the lost food crops. In South America, peasants and environmentalists who resist land grabs and hydroelectric projects are sometimes targets of paramilitary death squads. None of these effects are visible in the spreadsheets that inform climate policymakers. What we don't count, we don't know.

But at least the biofuels result in less atmospheric CO_2, right? Well, not necessarily. It depends on how you do the math. Do you include the lost carbon sequestration potential of the ruined ecosystem? The carbon released by increased levels of soil erosion? The unpredictable effects of disruptions in the hydrological cycle? The effects of local farmers leaving the land for the cities, where they become consumers in the global food system? Ignore these and you will be able to maintain the belief that bio-fuels are a fine thing for the planet. No doubt, that is what the biofuels companies believe. These people are not evil; they, like most of us, live in a story that valorizes their choices. That is why we need to propagate a new story that values people and place, soil and water, biodiversity and life; the qualitative and the relational.

Climate arguments have also been invoked in favor of giant woodchip-ping operations that are destroying forests in the southeastern United States and Eastern Europe. Close examination reveals these arguments as bogus, but when the policy establishment is in the habit of trusting the numbers, it is vulnerable to biased numbers, especially when financial interests of politically powerful lobbies are involved. And so, enormous woodchipping machines lower their hoods over one treetop after another, roaring down upon each and, in a matter of seconds, converting a living being into "climate-friendly biofuel."[5]

The problem here is not with biofuels per se. The problem, as with many other technologies, comes mostly with industrial scale and blind-ness to local ecological effects of production. Similarly, we adopt photo-voltaic and wind energy in the name of environmental health, counting the tons of carbon they replace while ignoring the toxic waste produced in PV panel and lithium-ion battery manufacture, and the birds and bats wind turbines kill. Those who bring up such issues are marginalized as nitpick-ing naysayers. All the more invisible are issues like adverse health effects from wind turbine noise (and who knows the effects of noise on wildlife?) or the climate consequences of what one indigenous person called "steal-ing the wind." What we don't know, we don't count.

To those wedded to the quantitative approach to problem-solving, any failure of quantification is to be remedied with even more quantification. Metrics-based thinking says that to remedy abuses of metrics, we need to extend them further, so that our measurements accurately encompass the uncounted emissions and lost sequestration. If only we could extend our

---

[5]   For an impressive video of these machines in action, see Blocker (2014).

measurements to totality, we would be able to make optimum decisions. But will our measurements ever be complete? No. Something will always be left out—the image of what we devalue.

What is typically measured is that which serves the economic and political interests, and unconscious biases, of those who commission the measurements. Then there are those things that we don't bother to measure because they are fundamentally unmeasurable, such as the sacredness of land, or of the water feeding the Ganges. Other cultures might say this river, that mountain, this forest is sacred. Is this just superstitious thinking that gets in the way of rational decision-making? Considering that our culture is ruining the planet whereas others, that had a sense of the sacred, lived sustainably on it for thousands of years, perhaps we should be cautious about imposing the value system encoded in our measuring onto the world.

By focusing on a measurable quantity, we devalue that which we cannot measure or choose not to measure. Such issues such as biodiversity, toxic pollution, radioactive waste, etc., not to mention social injustice and economic inequality, recede in urgency under the regime of carbon accountancy. Certainly one can make carbon-based arguments on all these issues, but to do so is to step onto dangerous ground. By saying "Stop the cement plant because of CO_2," you are also implying "If CO_2 weren't a problem, it would be fine." Right off the bat, you eliminate as allies anyone who doesn't believe in climate change. If global warming falls out of scientific favor, then all the environmental arguments pinned to it would collapse as well.

Imagine that you are trying to stop a strip mine by citing the fuel use of the equipment and the lost carbon sink of the forest that needs to be cleared, and the mining company says, "Okay, we're going to do this in the most green way possible; we are going to fuel our bulldozers with biofuels, run our computers on solar power, and plant two trees for every tree we chop down." You get into a tangle of arithmetic, none of which touches the real reason you wanted to stop the mine—because you love that mountaintop, that forest, those waters that would be poisoned.

The failures of carbon-motivated policies have something in common— they emphasize the global over the local, the distant over the immediate, and the measurable over the qualitative. This oversight is part of a more general mentality that sacrifices what is precious, sacred, and immediate for a distant end. It is the mentality of instrumentalism that values other beings and the earth itself in terms of their utility for us; it is the hubris of believing we can predict and control the consequences of our actions; it is the trust in mathematical modeling that allows us to make decisions according to the numbers; it is the belief that we can identify a "cause"—a cause that is something and not everything—and that we can best understand reality by dissecting it and isolating variables.

Usually, making decisions "by the numbers" means making them according to financial considerations. Is it really a very deep change to take the same methods and mentality and apply them instead to some other number?

We are in familiar territory in addressing problems by attacking their isolable, direct causes. That again is the mentality of war—end crime by deterring the perpetrators, end evil by dominating the evildoers, end drug abuse by banning drugs, stop terrorism by killing the terrorists. But the world is more complicated than that. As the War on Crime, the War on Drugs, the War on Weeds, the War on Terrorism, and the War on Germs show us, causation is usually not linear. Crime, drugs, weeds, terrorism, and germs might be symptoms of a deeper, systemic disharmony. Poor soil invites weeds. A run-down body offers a salubrious environment for germs. Poverty breeds crime. Imperialism begets violent resistance. Alienation, hopelessness, loss of meaning, and disintegration of community foster drug addiction. To address the complex of deep causes is a lot more difficult than to find something to blame and attack it using the familiar reductionistic methods.

Climate change is the same. It is a symptomatic fever of a deeper disharmony, a disharmony that pervades all aspects of our civilization. The fundamentalist wants to reduce every thing to one thing. That is convenient, if you would rather not look at everything.

As with terrorism, drugs, or germs, if we crack down on the proximate cause without addressing the underlying condition, the symptoms will return in a new and more virulent form. Similarly, when we make decisions by the numbers, then that which is not measured, the excluded other, will come back to haunt us.

Earth is a complex living system whose homeostatic maintenance depends on the robust interaction of every living and nonliving subsystem. As I will argue later, the biggest threat to life on earth is not fossil fuel emissions, but the loss of forests, soil, wetlands, and marine ecosystems. Life maintains life. When these relationships break down, the results are unpredictable: global warming, perhaps, or global cooling, or the increasingly unstable gyrations of a system spinning out of control. This is the threat we face, and because it is multifactorial and nonlinear, it cannot be overcome by simply reducing CO_2 emissions.

The Social Climate

While most environmentalists also care deeply about social justice, environmental narratives and particularly the climate narrative often suggest that social issues are of secondary importance compared to the grand mission to save the planet. Earlier I observed that this call for sacrifice to fight an overarching threat is identical to the way war is used to override social justice movements. "Stop whining—don't you know there's a war on?" My associate Marie Goodwin once asked a prominent climate crusader, "But don't you think community-building is also important today?" He replied, "Not really. If we don't put everything we've got into stopping climate change right now, we won't have any community to build." This pattern of thinking that climate change shares with war, I said, should make us alert. Now I am making a further point. It is not that "social justice is important too" (leaving unsaid, "but not as important as saving the planet"). It is that social healing is indispensable in ecological healing.

First and most obviously, it is indispensable because it is hard to effectively enact love for others when one is hurting desperately oneself. A hurting person usually passes on that hurt to those they love, helplessly. Contrary to what the more fortunate among us might think, when the alcoholic man abuses his children it is not because he loves them less than we love our own children. As with a hurting person, so with a hurting society. We cannot expect a miserable, oppressed populace to exercise much care for anything outside its immediate survival and security. While the poor are kept in a state of survival anxiety through sheer deprivation, the rich suffer poverty of another kind: lack of community, connection, meaning, and intimacy, which can cause severe psychological stress even in conditions of material plenty.

Most human suffering on this planet comes not from unavoidable tragedies like accidents and natural disasters, but from human beings themselves. Human trafficking and sweatshop labor, political violence and domestic abuse, racial oppression and gender violence, poverty and war ... all co-arise with our systems, our perceptions, and our narratives. These narratives are born of trauma and give birth to trauma.

Herein lies a link between economic justice, social justice, and the environment. We will continue to abuse our fellow beings, even our own Mother Earth, as long as we carry unhealed social traumas. This does not mean "heal our traumas first before we try to heal the environment." It is to recognize that social healing and ecological healing are the same work. Neither is to be privileged over the other; neither can succeed without the other.

From the causal logic of interbeing—morphic resonance—it is easy to understand how a society that exploits and abuses its most vulnerable will also exploit and abuse nature. To take care of vulnerable people generates a field of care that facilitates care for other vulnerable beings. A caring society is one that is habituated to asking, "Who is being left out? Who is suffering? Who is unrecognized in their gifts? Whose needs are not being met?" These are the questions that must guide an ecological society as well as a just society.

The term "social justice" may be too narrow to encompass the kinds of social healing that must happen for us to be able to fully enact our love for the planet. Traditional areas of social activism that aim to address racism, poverty, inequality, misogyny, and so forth are important but they leave unchallenged key institutions like education, medicine, money, and property, often engaging them only in terms of equal access. It is a very tepid form of activism to strive for the equal application of existing systems, when the systems themselves are inherently oppressive whatever the race, gender, or sexual orientation of their subjects.

Is what feminists want, that women have equal representation in the ranks of polluters, vulture fund CEOs, sweatshop owners, and slumlords? Is what Black Lives Matter activists want, that the school-to-prison pipeline be open as widely to whites as to blacks in a punishment-oriented carceral society? I suppose if we take the current system for granted, then the answer would be yes. If we take for granted a horrible concentration of wealth, then certainly all races should have equal odds of being in the elite or the underclass. If we take for granted a global war machine, then I suppose women should be allowed to be generals just as men are. If we take for granted a planet-wrecking economy, then female, gay, black, disabled, and transgendered people should be just as welcome at its helm as white males.

While they surely would protest at the above suppositions, the liberal media upholds them implicitly by celebrating every time a female superhero gets to kick ass in a film, or a government appoints black, gay, or female people to high positions. But the original feminist and racial justice radicals had a bigger vision than winning equality in the existing system. The feminists didn't want just to have equal status in the patriarchy; they wanted to transform the whole system. Civil rights leaders like Malcolm X and Martin Luther King Jr. didn't just want African American men to be treated equally in the United States military; they wanted to end militarism and imperialism altogether. But today, a neutered mainstream version of both civil rights and feminism settles for an anodyne ideal of equality, shifting the occupants of our power structures around but leaving the structures themselves intact. They seem not to realize that these structures necessitate inequality, whether delineated by race, gender, or some other distinction. An exploitative system requires some people to be exploited. Racial prejudice, male chauvinism, nationalism, etc., enable and

anodyne - not offensive

justify such a system, but eliminating these forms of bigotry won't change the underlying dynamics. Someone else will be exploited instead.

I visit this issue for two reasons. First, I want to make clear that social justice must be more than the usual grab bag of identity politics issues. The kind of social healing we need requires the massive overhaul, probably the total reformation, of our systems of medicine, education, birth, death, law, money, and government. Second, the same pattern of reaching for superficial changes that don't disturb the underlying system afflicts environmentalism just as much as it does social justice. So, just as a company can hire black, female, and LBGTQ executives at headquarters to administer a supply chain that exploits dark-skinned people in overseas factories and believe itself to be progressive, so also can it offset its carbon emissions by paying into a reforestation fund, all the while sourcing environmentally toxic products, and still call itself green.

The point is not to condemn the green rationalizations of corporations (or you or me); it is to illuminate the mindset of fundamentalism that enables those rationalizations. Fundamentalism of all kinds is a disengagement from the complexity of the real world, and I am afraid it is ascendant in many realms, not only religion. I even see it in various theories in alternative medicine, in the form of the Great Revelation of the One True Cause of all disease. (It's parasites! Inflammation! Stress! Acidosis! Trauma!) Fundamentalism offers certainty, a lockdown of thought into a few prescribed pathways. The rush to The Cause, the retreat to unquestioned axioms taken on faith, does not serve us in a time of the disintegration of so much of what we thought we knew.

If we continue to uphold climate fundamentalism, then the symptomatic fever of climate change will only worsen, whatever macroscopic measures we take to address its proximate causes. We might get the numbers down (temperature, greenhouse gases); yet, like the patient who goes to the doctor and is told, "The tests all say you are healthy," the disease will emerge in what we choose not to measure, what we cannot measure, or what is unmeasurable. We have to go beneath symptoms, and restore the foundation of ecological health: the soil, the water, the trees, and fungi, the bacteria, and every species and ecosystem and human culture on earth.

The Rush to a Cause

"That estuary used to be full of kelp and eels when we were kids," said Stella. "It was full of all kinds of wildlife. Crabs, clams, horseshoe crabs—there was a mussel bed right over there—one time I was swimming in that bend and came face to face with an eel."

That is what my wife told me on a visit to the estuary where the Narrow River meets the Narragansett Bay in Rhode Island, one of her haunts when she was growing up. It is a picturesque spot, surrounded by trees and sandy beaches, and I wouldn't have known that it is a seriously degraded ecosystem if Stella hadn't told me what it was like when she was a child.

Neither of us knows the reason why the eels disappeared. We shared a moment of sadness, and then Stella recalled another memory that somehow seemed to explain it. She and her friend Beverly would sometimes visit that part of the beach on what they called "rescue missions." Groups of marauding boys would come and flip over all the horseshoe crabs that had crawled onto the sand, leaving them to die there helplessly. Stella and Beverly would flip them rightside-up again. "Whoever was doing it had no reason to whatsoever," she said. "It was senseless killing."

This is the kind of story that makes me feel like I've detoured onto the wrong planet.

We didn't see any horseshoe crabs on this visit. They are a rare sight here now. I don't know if that is because people killed too many of them, or because we are "harvesting" too much of their blood for hemocyanin. Or maybe it is because of the general deterioration of the ecosystem, or pesticide runoff, agricultural runoff, land development, pharmaceutical residues, changing patterns of rainfall caused by development or climate change.... Maybe the horseshoe crabs are sensitive to one of these, or maybe the creatures they eat are, or it could be that the sensitive one is a mollusk that serves a role in the reproductive cycle of a microorganism that keeps another microorganism in check that infects the horseshoe crab.

I feel quite sure that whatever the scientific explanation for the die-off of the horseshoe crabs and eels, the real reason is the senseless killing Stella described. I mean not so much the killing part, but the senseless part—the paralysis of our sensing function and the atrophy of our empathy. We feel not what we do.

The crabs and kelp and eels are all gone. The mind searches for the cause—to understand, to blame, and then to fix—but in a complex nonlinear system, it is often impossible to isolate causes.

This quality of complex systems collides with our culture's general approach to problem-solving, which is first to identify the cause, the culprit, the germ, the pest, the bad guy, the disease, the wrong idea, or the bad personal quality, and second to dominate, defeat, or destroy that culprit. Problem: crime; solution: lock up the criminals. Problem: terrorist acts; solution: kill the terrorists. Problem: immigration; solution: keep out the immigrants. Problem: Lyme Disease; solution: identify the pathogen and find a way to kill it. Problem: racism; solution: shame the racists and illegalize racist acts. Problem: ignorance; solution: education. Problem: gun violence; solution: control guns. Problem: climate change; solution: reduce carbon emissions. Problem: obesity; solution: eat less and burn more calories.

You can see from the above examples how reductionistic thinking pervades the entire political spectrum, or certainly mainstream liberalism and conservatism. When no proximate cause is obvious, we tend to feel uncomfortable, often to the extent of finding some reasonable candidate for "the cause" and going to war against that. The recent spate of mass shootings in America is a case in point. Liberals blame guns and advocate gun control; conservatives blame Islam, immigrants, or Black Lives Matter and advocate crackdowns on those. And of course, both sides especially like to blame each other.

Superficially it is obvious that you can't have mass shootings without guns, just as it is obvious that allowing civilian access to military assault weapons makes those shootings more deadly. However, to focus attention on gun availability bypasses more troubling questions that don't admit easy solutions. Where does all that hatred and rage come from? What social conditions give rise to it? The furious debate about gun control so monopolizes U.S. political attention that these questions languish on the intellectual margins. If we do not address them, does taking away the guns really do much good? Someone could use a bomb, a truck, poison ... is the solution then a complete lockdown of society, the imposing of ubiquitous and ever-increasing surveillance, security, and control? That is the solution we've been pursuing my whole lifetime, but I haven't noticed people feeling any safer.

Perhaps what we are facing in the multiple crises converging upon us is a breakdown in our basic problem-solving strategy, which itself rests on the deeper narratives of the Story of Separation.

Learning of the die-off of the estuary, I myself felt the impulse to find the culprit, to find someone to hate and something to blame. I wish solving our problems were that easy! If we could identify one thing as *the* cause, the solution would be so much more accessible. But what is comfortable is not always true. What if the cause is a thousand interrelated things that implicate all of us and how we live? What if it is something so all-encompassing and so intertwined with life as we know it, that when we glimpse its enormity we know not what to do?

That moment of humble, powerless unknowing, where the sadness of an ongoing loss washes through us and we cannot escape into facile solutioneering, is a powerful and necessary moment. It has the power to reach into us deeply enough to wipe away frozen ways of seeing and ingrained patterns of response. It gives us fresh eyes, and it loosens the tentacles of fear that hold us in normality. The ready solution can be like a narcotic, diverting attention from the pain without healing the wound.

You may have noticed this narcotic effect, the quick escape into "Let's do something about it." Of course, in those instances where cause and effect is simple and we know exactly what to do, then the quick escape is the right one. If you have a splinter in your foot, remove the splinter. But most situations are more complicated than that, including the ecological crisis on this planet. In those cases, the habit of rushing to the most convenient, superficially obvious causal agent distracts us from a more meaningful response. It prevents us from looking underneath, and underneath, and underneath.

What is underneath the callous cruelty of those horseshoe crab flippers? What is underneath the massive use of lawn chemicals? What is underneath the huge suburban McMansions? The system of chemical agriculture? The overfishing of the coastal waters? We get to the foundational systems, stories, and psychologies of our civilization.

Am I saying never to take direct action because after all, the systemic roots are unfathomably deep? No. Where the unknowing, perplexity, and grief take us is to a place where we can act on multiple levels simultaneously because we see each dimension of cause within a bigger picture and we don't jump to easy, false solutions.

The Mother of All Causes

When I wondered about the cause of the estuary die-off, a hypothesis may have jumped into your mind—climate change, the culprit *du jour* for nearly every environmental problem. *If we could identify one thing as the cause, the solution would be so much more accessible.*

As I was doing research for this book, I googled "effect of soil erosion on climate change," and the first ten pages of results showed the converse of my search—the effect of climate change on soil erosion. I did the same for biodiversity, with the same result. Whether or not it is true that climate change exacerbates other environmental problems, the rush to name a unitary cause of a complex problem should give us pause. The pattern is familiar. It is none other than war thinking, which also depends on identifying a unitary cause of a complex problem. That cause is called the enemy, and the solution is to defeat that enemy.

Carbon reductionism is like "germ reductionism" in medicine. What is the cause of, say, strep throat? Well, obviously it is the streptococcus bacterium, right? Problem: germ; solution: kill the germ. On one level this may be accurate, but consider what this approach renders invisible and leaves out. First, it leaves out the question why one person exposed to the germ gets sick, and another does not. Especially if someone gets repeated infections of strep, it might be more useful to see the germ not as the cause, but as one of the symptoms of the disease. It also ignores the effects of repeated antibiotic treatment, and whether that might somehow contribute to vulnerability to reinfection. (This is not idle speculation, in light of recent science establishing the relationship between body ecology and immunity. Body ecology—which includes a healthy gut microbiome— is seriously disrupted by antibiotics.)

In medicine, focusing on the immediate, linear cause of a disease can vitiate the possibility of a real cure, whether on an individual or epidemiological level. Consider a disease that looms far larger than strep in our collective consciousness today: Lyme Disease. Seeing it as an infection by a tick-borne spirochete, the appropriate technologies of control are obvious: avoid or destroy ticks, and kill the spirochete. To see the problem in another way can be very inconvenient or disruptive to the systems that embed the usual control-based responses. What is the real "cause" of Lyme? I don't know, but it could include the following:

- Compromised immunity that leaves the body vulnerable to numerous viral coinfections (against which antibiotics are useless)

- The explosion of deer tick populations due to increasing "edgelands" and loss of deep forest caused by suburban development

- Exploding deer populations caused by the extermination of wolves and cougars

- Declining forest health and understory degradation caused by pollution, repeated clear-cutting, and again predator extermination (deer destroy the forest understory when their numbers are unchecked by predation), which in turn devastates biodiversity and allows overgrowth of species like ticks

- Decline of pheasant and other tick-eating birds, because of historical overhunting, roadkill, and understory destruction

- Widespread aerial spraying of insecticides to control gypsy moth caterpillars and other insects, decimating insect-eating birds

- The exteriorization of modern fear of nature; it is as if nature responds to our locking children in the safety of indoors by saying, "Okay, I'll give you something to be afraid of for real."

We can dig deeper and deeper. What is the cause of suburbanization? What is the cause of pollution? What is the mentality behind the extermination of top predators and the spraying of forests? Complex, nonlinear causal relationships connect these phenomena. For instance, the othering of nature encourages suburbanization, but it is also the other way around. Cut off from direct relationship to the land, the suburbanite who buys food grown thousands of miles away and needn't even tread the soil to move from place to place will of course see nature as a spectacle or a threat.

One might even say that "the cause" of Lyme is everything and anything. Even the locution "the cause" is part of the problem, as it implies the separability of interdependently co-arising phenomena. I could even say the cause of Lyme Disease is modern children's storybooks that present us from a young age with anthropomorphized animals wearing human clothes, living modern lives, and thinking human thoughts. Such storybooks entice us to take other beings on our terms and not theirs, and obscure the fact that the human normality that the storybook animals are playacting is extinguishing the actual habitats of those animals in the real world.

Now I am not saying that one should never address an obvious, linear cause, just as I am not saying that there is never a time for a fight. I am warning, rather, of the habit and conditioned response of addressing all problems in this way.

In ecology, which is the study of relationships and not things, every cause is also a symptom. Let's take for example the steep decline of seagrass meadows, which are biodiversity hotspots and sequester more carbon per hectare than nearly any other ecosystem. Seagrass die-off is a cause of carbon loss and acidification, and it is a symptom of:

- Proliferation of herbivorous mollusks and crustaceans, caused by the overfishing of larger predatory fish

- Eutrophication and algae blooms, caused by excess agricultural runoff

- High levels of silt that reduces seagrass's access to sunlight, which is a consequence of soil erosion, which is a consequence of modern agricultural practices, logging, and development

According to a friend who works with "watermen" (mostly crabbers) on the Chesapeake Bay, seagrass, shellfish, and crabs experience a massive die-off every time there is a hurricane or severe influx of highly sedimented freshwater into the bay. These irregular disruptions keep the ecosystems ever precarious. Well, that wasn't much of a problem a few centuries ago, because:

- Intact wetlands could absorb massive amounts of rainfall.

- Beaver dams along all the small tributaries to the bay slowed runoff and trapped sediment.

- Deforestation and tillage had not exposed bare topsoil to erosion.

Clearly, protecting and restoring seagrass is more than a matter of roping off protected areas, because seagrass is in relationship with everything that lives, including ourselves. Nor will our normal find-an-enemy strategy save the seagrass. It is tempting and convenient to blame the problem on "more intense hurricanes caused by global warming," ignoring the complex of causes that intimately involves ourselves and the way we live. It is also easier to blame the watermen for their supposed greed, ignoring the complex economic causes (in which, again, we all participate) that drive the relentless conversion of nature into commodity into money. Our intellectual habit is to find the One Cause, our scientific programming is to measure it, and our political gearing is to attack it. When the One Cause is global, we cross our fingers and hand over responsibility and power to distant global institutions. They'll take care of it. We hope. But too often, blaming climate change means not doing anything at all.

Like most binary distinctions, that between symptom and cause breaks down under close scrutiny. Yet the distinction is still useful. Causes are symptoms and symptoms are causes, yes. So let us name as "symptom" that aspect of the cause/symptom complex that presents itself most obviously to our attention. To us, Lyme is calling most loudly. To another culture, the most alarming change might be the disappearance of the dogwoods in the mid-Atlantic forests, or perhaps some change in the songs of birds that you and I would never notice. Thus, what we observe to be happening in the world says as much about ourselves as it does about the world. It reveals what we think is important, significant, valuable, and sacred, and what is irrelevant or useless too. Put another way, what we see reveals how we see.

Nerdy aside: I am not in this book (or anywhere) taking the postmodern position that reality and truth are human cultural constructions—that how we see is the only determinant of what we see, or that there is no is-ness outside human seeing. Maybe the postmodern philosophers are right that there are no facts, but only meanings loaded with power dynamics, gender and racial oppression, etc. But what they cannot countenance is that we humans are not the only meaning-makers, not the only authors, not the only full subjective agents. Our ways of seeing, our stories and our myths, come from a source beyond our comprehension.

Among the many causal narratives available to apprehend Lyme, or climate change, or any other issue, our culture chooses the one that best preserves the status quo. The dominant culture adopts the narrative that sustains its dominance.

People tend to conceptualize problems in such a way as to validate the tools that are familiar and available to them. If all you have is a hammer, everything looks like a nail. If all you have are antibiotics, you will always look for the germ. If all you have is the mindset of war, then you will always look first for an enemy.

Our society's most potent and familiar tool is the quantitative methods of science. That is therefore how we frame the problem of climate change. We use numbers (such as average global temperatures) to prove it is happening, other numbers (CO_2 emissions) to formulate responses, and yet other numbers (embedded in computer modeling) to forecast the future and guide policy. But is this the only tool? Is this even the right tool? We might doubt it, given the damage industrial civilization has caused the planet depends on the same regime of quantification. Through science we describe the world in numbers and mathematical relationships. Through technology we apply those numbers to the control of the material world. Through industry we convert the world into commodities, characterized by numerical specifications. Through economics we further convert all things into another number called its value.

We would like to solve climate change with methods and mindsets that are familiar to us, for to do so would preserve the foundation of society as we know it. These methods and mindsets, the quantified worldview, tell us that we can fix the situation by eliminating fossil fuels. Unfortunately, as I will discuss later, the mere elimination of fossil fuels will not deliver us from the ecological crisis. A deeper revolution is afoot.

Eliminating fossil fuels does not represent as thorough a change as the change required to halt ecocide here, there, and everywhere. Conceivably, we could eliminate carbon emissions by finding alternative fuel sources to power industrial civilization. It may be unrealistic upon deeper investigation, but it is at least conceivable that our basic way of life could continue more or less unchanged. Not so for ecosystem destruction generally, which implicates everything upon which modern technological society depends: mines, quarries, agricultural chemicals, pharmaceuticals, military technology, global transport, electronics, telecommunications, and so on. All of these must evolve into their next incarnation; some may even become obsolete.

The Place Where Commitment Lives

The equation of "green" with "low carbon," which maps a complex matrix of causes onto a single quantifiable variable, leads us to think that sustainability can mean to sustain life as we know it, life as it has been. It justifies and motivates the operating paradigms of green growth and sustainable development, which are essential to preserving our present economic system with its endless appetite for more resources. For that matter, it allows us to continue seeing the planet as composed of said "resources"—things that are here for us to use—as long as we exploit them in a way that doesn't generate greenhouse gases. And, crucially, it contributes to an attitude of humanity in the driver's seat, managing planet Earth like a machine, controlling the inputs and measuring the outputs. It invites a linear response to a nonlinear problem. But Earth is not a machine; it is alive, and it will remain hospitable to life only if we treat it as such.

In coming chapters I will present evidence that the climate effects of deforestation, industrial agriculture, wetlands destruction, biodiversity loss, overfishing, and other maltreatment of land and sea are far greater than most scientists had believed; by the same token, the capacity of intact ecosystems to modulate climate is much greater than had been appreciated. This means that even if we cut carbon emissions to zero, if we don't also reverse ongoing ecocide on the local level everywhere, the climate will still die a death of a million cuts.

Contrary to the presupposition implied in my aforementioned Google search results, the health of the global depends on the health of the local. The most important global policies would be those that create conditions where we can restore and protect millions of local ecosystems. Today it is often the opposite; for example, global free trade treaties permit corporations to sue governments for lost profits from local environmental protections.

When we cast ecological healing in global terms, our gaze wanders away from the places we have loved and lost, the places that are sick and dying, the places we care about that are tangible and experientially known and real to us. It goes instead to distant times and places, and our local loves become at best instruments toward a larger end.

Why was Stella sad to see her beloved estuary depleted of life? Was it that it no longer grows kelp that will sequester carbon and mitigate climate change? Of course not. If so, it would be no great loss. It could be offset by planting a kelp farm or a forest somewhere else, or perhaps by installing giant carbon-sucking machines in every city. Then Stella would be happy, right?

My friend Seppi Garrett told me how he took his son fishing in the Conodoguinet Creek, his favorite haunt as a boy. To his alarm, he found out that the creek is impaired and people are warned not to let children get into the water. So he thought, "I'll take him to the Yellow Breeches river instead," only to discover that it is impaired as well. He said, "Then of course there is the Susquehanna. I feel so sad when I go there and see oil slicks on the water in the places I used to wade chest-deep to go fishing when I was a kid." Seppi's grief, indignation, and anger are driving him to become a kind of freelance applied ecologist, part of a movement of people who assist the recovery of damaged areas by accelerating succession, redirecting water, and altering species composition. We need millions of people to do that, to listen closely to land, to develop a relationship with it, and to put themselves at its service. Where does that level of commitment come from? Again I will ask, does the oil slick provoke Seppi's grief because it signifies fossil fuel burning that generates carbon dioxide?

You can see how the dominant global carbon narrative is not necessary to generate environmental zeal, even for those who accept it as true. All the more, for the climate change skeptics we'll visit in the next chapter.

I am sure something stirred in you reading Seppi's words, even if your own special childhood place was the woods not the river. When we transmit our love of earth, mountain, water, and sea to others, and stir the grief over what has been lost; when we hold ourselves and others in the rawness of loss without jumping right away to reflexive postures of solution and blame, we are penetrated deep to the place where commitment lives.

This does not mean we don't face a global ecological crisis. We do, and it far transcends what we call climate change. However, if everyone focused their love, care, and commitment on protecting and regenerating their local places, while respecting the local places of others, then a side effect would be the resolution of the climate crisis. If we strove to heal and protect every estuary, every forest, every wetlands, every piece of damaged and desertified land, every coral reef, every lake, and every mountain, not only would most drilling, fracking, and pipelining have to stop, but the biosphere would become far more resilient too.

3

The Climate
Spectrum and Beyond

Which Side Am I On?

The foregoing critique of the dominant climate change narrative may have the reader wondering which side I am on. That is always the most important question in a war. Do I, in spite of my critique of reductionism, still affirm the basic principle that carbon emissions pose a grave and immediate threat to the climate? Or am I, instead, a "climate change denier"? Which side am I on in the "fight" against climate change?

As I elaborate the critique I began in the last chapter, it will become clear that this is the wrong question: wrong in its emphasis, wrong in its implications, and wrong in the worldview that underlies it. For now I will say that this book takes a position that is both skeptical and alarmist. It is skeptical of certain aspects of the dominant narrative of climate change, while affirming that human activity is alarmingly destabilizing the ecosphere. If anything, I tend toward an extreme view of the gravity of the ecological crisis. The prescriptions herein partially align with conventional climate advocacy, and in some respects far exceed it, albeit for different reasons and from different motivation. I hope, therefore, my arguments will be persuasive even to those who disbelieve in anthropogenic global warming (AGW). For AGW believers, this book may offer new political and material strategies for addressing climate change as part of a broader ecological regeneration.

What you will see as I deconstruct the conventional spectrum of opinion on climate change is that the dynamics of the debate obscure something more important than which side is right. As with many polarizing issues, it is the hidden assumptions, shared by both sides and questioned by neither, that are most significant and most potent in taking us into new territory.

These hidden assumptions include agreements about what is significant as well as agreements on what not to talk about. To give an example from another realm, in the political debate on immigration, one side says to keep them out, the other side says let them in, and eventually governments institute a policy somewhere in the middle. But neither side asks, "What are the policies that make life in other places so unlivable that people risk their lives and separate from their families to emigrate?" Both sides agree not to talk about military imperialism, neoliberal trade policies, and the global debt regime, or are not even conscious of them. Yet without a change on that level, the immigration issue will never be resolved. The furious mainstream debate draws all the attention away from the underlying causes and toward the superficial symptoms. Therefore, it perpetuates the status quo.

Most polarized conversations are like this, whether in politics or within communities or couples. They are part of a holding pattern that absorbs and squanders the energy of discontent, leaving the real issue untouched. Usually, the real issue is more uncomfortable, because it involves not only the demonized opponent, but oneself as well.

Here is a map of the conventional spectrum of opinion on climate change, which as you will see includes positions that seem extreme and mutually opposed. They are not. However acutely opposed they may seem, hidden agreements unite them, and it is these hidden agreements that make the problem unsolvable. Where we need to go, and where the ecological crisis will eventually take us, is off this spectrum entirely.

1. **Climate change skepticism.** This says either that climate change, especially global warming, is not happening, or that if it is happening it has little to do with human activity, or that if it is attributable to human activity it isn't dangerous. Sometimes all three positions appear on the same skeptic website. Recently, even modest departures from climate orthodoxy draw the epithet "denialism."

2. **Techno-optimism.** Climate change is another of the challenges we will overcome in the triumphant onward march of technology. Through geoengineering and alternative energy technologies, we will bring down greenhouse gas levels and find other ways to cool the atmosphere. Human creativity is limitless; there is no problem we can't solve if we put our minds to it. What is needed is simply to turn our focus to these problems, to incentivize their solutions, and thereby to unite science and finance toward the meeting of humanity's newest challenge.

3. **Climate orthodoxy.** The burning of fossil fuels poses a grave threat to humanity and the planet. If we do not act quickly to cut emissions and limit warming to 2°C, the future will bring rising sea levels; extreme weather, floods, and droughts; crop failures; famine and mass migrations; and devastation to marine and land ecosystems. Therefore, we need to wean ourselves off fossil fuels as quickly as possible, transition to carbon-neutral energy technologies, and encourage economic policies of sustainable development and green growth. The window for action is shrinking; there is no time to waste.

4. **Climate justice and systems change.** This viewpoint is a step toward a deeper radicalism. It says that climate change is inextricably linked to our economic system and various institutions of social oppression. Climate change isn't only an environmental issue, it is a social, racial, and economic issue; moreover, the system's dependency on the profits from a fossil-fuel-based industrial economy

means that climate change can be addressed only by changing capitalism as we know it.

5. **Climate catastrophism.** This viewpoint basically says that it is already too late to prevent catastrophic climate change, except perhaps with an immediate response far beyond anything that is politically conceivable today (and perhaps not even then). The more moderate prognosticators foresee a dramatic collapse of society: a population crash, sociopolitical upheaval, and a major regress in technological levels. Those on the more extreme end predict temperature rises of 6–10°C within the next few decades, which would mean the end of civilization and possibly the extinction of the human species. Some such as Guy McPherson predict human extinction within ten years.[1]

What could possibly unite such disparate viewpoints? First, they share a focus on greenhouse gases and global temperatures. On one end of the spectrum, it is that they aren't a problem; on the other, that they spell the end of civilization. All agree with the general consensus that puts climate change and carbon at the heart of environmentalism.

Accordingly, the skeptics (most of them anyway, but not all) throw out the baby of care for nature with the bathwater of the standard AGW narrative. Similarly, alarmists unwittingly demote other environmental issues (not to mention social issues) to secondary importance in their focus on AGW.

The intensity and ubiquity of the conversation around this issue sucks the air out of the room for issues like wildlife conservation, habitat preservation, toxic and nuclear waste, soil erosion, aquifer depletion, and so on. Tragically, as I will argue, it is precisely these other issues that are the hidden drivers of climate instability. Climate change is a symptom of ecosystem degradation, a process that goes back at least five thousand years and has reached peak intensity today. It arises from the basic relationship that has prevailed between civilization and nature.

---

[1]   Early in 2017 I heard him predicting with confidence human extinction in two to four years.

Climate change is inviting us to forge a different kind of relationship, one that holds the planet and all of its places, ecosystems, and species sacred—not only in our conception and philosophy, but in our material relationship. Nothing less will deliver us from the environmental crisis that we face. Specifically, we need to turn our primary attention toward healing soil, water, and biodiversity, region by region and place by place. Endless photovoltaic arrays on ruined land are not going to solve the problem. We must enact a civilization-wide unifying purpose: to restore beauty, health, and life to all that has suffered during the Ascent of Humanity.

Across the spectrum, carbon dominates the conversation. Most (but again, not all) skeptics seem to want the environmental problem to go away altogether, and hope that by refuting climate change we will once again have unlimited license to pillage the planet. The climate fundamentalists, despite their general sympathy for other environmental causes, instigate a perversely similar banishment of other environmental issues that gives implicit license to any sort of ecological pillage that doesn't generate CO_2.

I am suggesting here that the frame of the debate is itself part of the problem. The "frame of the debate"—drawing from the Story of Separation—includes:

- A conception of nature as "environment" and thus separate from ourselves

- The assumption that climate is governed primarily by global geomechanical processes (solar radiation, atmospheric gases, Earth's rotation, polar/equatorial heat differentials, etc.) and not by life processes

- A mechanistic view of nature as an incredibly complicated machine

- The primacy of a quantitative approach to knowledge

- Valuing other beings based on instrumental utilitarianism—their use-value to ourselves

- The belief that human beings are the only fully conscious, subjective agents on this planet

In overt and subtle ways, these assumptions inform climate science and policy today, from the formulation of basic research questions, to the political arguments about climate, to priorities in funding, technology, agriculture, and industry. They are shared by alarmists and skeptics alike, which is not surprising since the same assumptions also underpin civilization as we know it. The problem and the current modes of solution come from the same place. That is why a different framing is needed.

To put it in more shocking terms, it doesn't matter if the skeptics are right or not, because the assumptions on which the debate is based are already enough to doom us to a dystopian future. I would like therefore to offer a new "frame of the debate":

- Earth is a living organism.

- Each biome, local ecosystem, and species contributes in unique ways to the health and resiliency of the whole; they are the organs and tissues of the Gaian organism.

- All beings—plants and animals, soil, rivers, oceans, mountains, forests, etc.—deserve respect as alive, sentient subjects and not mere things.

- Any damage to the integrity of the planet or the beings on it *inevitably* damages human beings as well, whether or not the causal pathways for that damage are visible.

- Similarly, a healthy planet will benefit the physical and spiritual health of humanity.

- The psychic climate comprising our beliefs, relationships, and myths is intimately connected to the atmospheric climate.

- Likewise, the political climate and social climate are co-resonant with the atmospheric climate.

- The purpose of humanity is to contribute our gifts to the beauty, aliveness, and evolution of Earth.

The converging crises of our time, including the ecological crisis, are an initiation for our civilization. The belief system I just outlined awaits us on the other side of that initiation.

Can you imagine what a society would look like that embodied these beliefs in its agriculture, technology, and economics? Current "green" policies would seem paltry in comparison. Today, the policy ship of environmentalism must sail against the current of the Story of Separation. Pulling the oars furiously, the environmental movement stirs up a mighty froth, yet for all its progress through the water, the ship is carried backward by the current; the overall condition of the planet continues to worsen. Fifty years after the Clean Air Act, pollution planetwide is worse than ever. Forty years after the Clean Water Act, the ocean's plastic outweighs its fish. Forty years after the Endangered Species Act, biodiversity on earth is in precipitous decline. And after several decades of climate accords, climate derangement continues to intensify.

Is the solution to pull even harder on the oars? If the current is unchangeable, that would be the only hope. Here is where the metaphor breaks down, because the current is not some arbitrary force of nature or human nature, as if we were genetically disposed to destroy the world. No, the current is composed of systems created by human beings: first and foremost the financial system, and also our systems of government, science, technology, education, and religion. What human beings have created, they can uncreate.

How to uncreate them is no trivial matter. We should be skeptical of save-the-world narratives; historically, such quests have done more harm than good. Inevitably, and especially when they demand urgent action, they draw from the existing ingredients at our disposal: existing institutions of political power, existing economic mechanisms, existing modes of technology, and existing ways of thinking. To organize quick action on a large scale usually involves giving more power to institutions that wield power already. We need to look beyond existing institutions, ways of thinking, technologies, and economic mechanisms, all of which are intrinsic to the problem. Uncertainty lies ahead, new social territory in which we will discover unsuspected modes and expressions of human creativity.

I can, however, offer a guiding principle. Our system moves according to a deeper current still; namely, our civilizational mythology: the stories, meanings, perceptions, and agreements that constitute what we think to be reality. The world's healing must and will come from outside the mythology of Separation that brought us to the present impasse.

A Visit to the World of Skepticism

The us-versus-them drama that our culture seems automatically to reenact appears not only as the "fight against climate change" but also, in the search for an identifiable enemy, as a battle against those who doubt or deny that climate change is real. The thinking goes as follows: if only the unholy alliance among fossil fuel companies, their financiers and investors, their political allies, and a small minority of venal academics could be overcome, we would be able to take meaningful, swift action to halt climate change. The identity of the enemy is clear. We can settle in to the familiar operating framework of the fight.

A nearly universal tactic in warfare is the dehumanization of the enemy. Accordingly, the Standard Narrative of climate change activism says that those who disbelieve in anthropogenic climate change must not be in full possession of their mental or moral faculties. They are greedy, they are corrupt, they are delusional, they are in denial; they are hypocrites, liars, and psychopaths. Otherwise, how could they ignore the overwhelming evidence, the "settled science," the consensus of "97 percent of climate scientists"? It seems inconceivable and outrageous.

Trusting that I myself am not a hypocrite, liar, or psychopath, and am in possession of at least some fraction of my mental and moral faculties, I decided to explore the views of climate skeptics more deeply.

The climate skeptic camp turns the above accusations around and speaks of the incompetence and corruption of mainstream climate scientists. (Its more sophisticated adherents emphasize groupthink, publishing and funding bias, and political pressure as the main mechanisms by which orthodoxy is enforced.) In response to the label "climate denialism" they name the mainstream "climate alarmism."

It may seem from the above that I am leaning toward the side of the skeptics in drawing what may look to the believer like a false equivalency. After all, in World War II the Nazis and the Allies demonized each other as well, but that doesn't make the two sides equivalent. There were good guys and bad guys in that war (right?); all the more so in this one, where the survival of humanity is at stake.[2] To hint at the possible legitimacy of the enemy's position or to criticize the rationale for war is already an act of betrayal—"rendering aid and comfort to the enemy" it was called during the Bush administration's War on Terror. Likewise, it is an act of betrayal not to take sides. Such is the mentality of war.

In wartime, pacifists draw more hostility and contempt than the enemy does. Why? Because the pacifist questions the validity of the roles people identify with and the story they live in. They pose an existential threat— not to survival, but to identity.

In my exploration of the skeptic position, I adopted a kind of deliberate naïveté, rejecting *both* sides' characterization of the other and temporarily assuming that most parties to the debate are, albeit imperfectly, competent, intelligent, and sincere. I chose various of the main lines of the standard climate narrative and then read extensively the best skeptical blogs and websites I could find, to see what they actually say about what seems to be overwhelming evidence for global warming. I also read the best and most patient rebuttals I could find of the skeptics' arguments. Let me share a representative sample of my adventure, with my responses suitably exaggerated for dramatic effect.

I started with what looks like incontrovertible proof of AGW (anthropogenic global warming): Michael Mann's "hockey stick" graph showing

---

2   I'm quoting the standard narrative of World War II here. In reality, while there were clearly bad guys, it is not so clear that there were any good guys. The war against the Axis powers was inextricably tied in its historical origins and execution to American imperial ambitions; the defeat of even worse imperial powers was a happy side effect.

a rapid acceleration in global temperature in the twentieth century. In the graph, centuries of relatively stable temperatures precede a rapid warming closely congruent to the increase in atmospheric CO_2. You can't argue with the numbers. Certainly, correlation does not prove causation, but the absence of any other explanation for such a drastic, unprecedented rise makes a causal link likely, particularly in light of the greenhouse effect of CO_2. How could an intelligent person sincerely doubt such strong evidence?

I decided to find out. The climate skeptics claim that there are serious flaws in the statistical methods used to construct the hockey stick graph.[3] They criticize both current and historical data as unreliable, incomplete, and heavily "adjusted" always with a bias toward demonstrating recent warming—old numbers adjusted lower, recent numbers adjusted higher. The tree ring proxy data, they say, doesn't take into account that slower tree growth might be due to less CO_2 or less rainfall, not colder temperatures.[4] Current data they also claim to be unreliable due to urban heat island effects—compared to the past, an inordinate number of weather stations are located near air conditioning vents, parking lots, airports, water treatment plants, and other heat sources.[5] Moreover, raw data is adjusted upward in a process called homogenization.[6] If one weather station is giving results that are out of line with neighboring stations, its data is homogenized under the assumption that it is subject to a malfunction or microclimatic influences—but usually, the skeptics say, the ones giving lower readings are adjusted upward, often in comparison to stations that are subject to the warmer temperatures resulting from the presence of buildings or asphalt. These problems have led some researchers to look at alternative temperature datasets gathered by satellites, which aren't

---

[3] See, for example, Muller (2004) and for a more polemic take, see Krüger (2013). For a simplified summary of the basic statistical criticism, see Moriarty (2010).

[4] Moriarty (2010).

[5] See Watts (2009).

[6] For a reasonable presentation of this contention, see Steele (2013).

subject to the vagaries of widely distributed surface temperature readings. After all, theoretical models of greenhouse effects predict warming of the entire troposphere. These alternative datasets, say the skeptics, agree closely with each other and show a much slower temperature rise than the surface temperature data upon which the recent part of the hockey stick is built. In any event, present temperatures are still lower than during the Medieval Warm Period, which is the subject of repeated attempts to revise out of existence. Furthermore, say the skeptics, historical carbon dioxide levels follow and do not precede temperature increases, and often are not correlated at all. Ice core reconstructions of CO_2 use data from which data points have been removed when they contradict the standard narrative, on the grounds that they must have been contaminated.

My goodness—how could I have been such a fool as to believe the party line peddled by Big Science? I'd been duped along with everyone else into believing the orthodoxy. How could I have been taken in?

Just to make sure, I'll look at what mainstream climatologists say in response. Hold on here—things are not as the skeptics claim. The critics of the hockey stick are using one or two insignificant errors to throw out the entire paper; besides, the errors were corrected in the 2008 version of the paper. Since the original paper was published, other peer-reviewed research using numerous other proxies has confirmed again and again that the last two decades are the hottest in two thousand years.[7] There are now many, many "hockey stick" reconstructions of Paleoclimate data, all more or less consistent with Michael Mann's.

As for the satellite data, the skeptics don't realize that the orbital decay of the satellites introduces a spurious cooling effect that would have to be corrected for. You can't trust the raw temperature readings. Second, temperature readings are also skewed by "diurnal drift." Third, the satellites aren't really measuring temperature; they are measuring microwaves emitted by atmospheric oxygen, which is only indirectly a function of temperature. Fourth, the charts I'd been looking at rely on weighted averages

---

[7] Moriarty (2010).

of various levels of the troposphere that are weighted in a way that might exaggerate cooling; moreover, data from different types of sensors must be combined and fitted to a single scale. In any case, scientists took the discrepancies seriously, but when they investigated the reasons and adjusted the data, the result was that satellite data matches surface temperature data and theoretical models quite closely. Moreover, there are actually five satellite datasets, and the skeptics always display the one that shows the least warming—even though that one correlates the least closely with weather balloon data, another source of troposphere temperature measurements.[8]

Historical CO_2 levels, says the mainstream, only appear to follow temperature rises because rising temperatures kick off a positive feedback cycle, amplifying what would otherwise be minor warming.

As for the heat island effect and the data adjustments, says the mainstream, these have been handled very scrupulously in order to remove distortions in the raw data.[9] Besides, rural and urban weather stations are consistent in the degree of warming they show.[10] The same goes for the carbon dioxide levels in ice cores. Scientists had very good scientific reasons for eliminating outlier data points that could not be correct, since there is no possible mechanism by which CO_2 could have been at those levels. To ignore the lengthy conversations within the community of scientists and issue an armchair opinion that they have connived to manipulate the results according to some preconceived "agenda" is an insult to the scientists and reveals a profound lack of understanding about how science is really done.

Wow, I'm sure glad I read these rebuttals by real scientists who aren't on the payroll of the fossil fuel industry before I let any climate denialism infiltrate this book. I'd nearly been taken in by the deniers. Who do I think I am, anyway, to imagine that I know better than the climate scientists who have spent decades studying the topic? How arrogant to think that in a

---

[8]   Foster (2016).

[9]   For a partial explanation of how this is done, see: Hausfather and Menn (2013).

[10]   Mothincarnate (2015).

couple weeks of doing "research" on the internet I could find some obvious way they are wrong, and that they lacked the brains or integrity to see. I feel ashamed to have doubted them.

In the interests of due diligence, I'll see if the skeptics respond. They do. The 2008 version of Mann's paper, they say, contains the same basic flaws as the original, and other "hockey stick" studies use the same problematic temperature proxies. They claim that the reason that rural weather stations show the same upward trend as urban stations is that while defined as rural, many are also subject to significant urbanization. They say that in fact, the orbital decay factor was corrected for twenty years ago and in any case affected only the lower troposphere readings, which are not at issue here. Diurnal drift has been corrected for as well. The microwave emissions are a better measure of temperature than the electronic resistance method used for surface recordings. The climate establishment is constantly "adjusting the data" every time it doesn't fit their narrative or models, each "adjustment," of course, being in the upward direction. The datasets that conform to the weather balloon readings and demonstrate greater warming do so because they include data from a satellite that was not corrected for calibration drift, and then adjust the data for diurnal drift according to a climate model rather than empirical data.[11]

It looks like I was taken in again, bamboozled by the authoritative-seeming dismissals of the minority position without really understanding the science behind it.

What becomes apparent in this back-and-forth is that in the end I am probably unable to make my choice of belief on purely evidentiary grounds. When I pursued the question of temperature readings a bit further, I got mired in a morass of technical minutiae about atmospheric physics, statistical methods, and so forth that I lack the scientific background to easily understand. Mind you, I am scientifically literate and have a degree in mathematics from Yale University. If I can't judge the issue on its merits, how can the average citizen? Moreover, as the disagreements among those who *do* have the scientific

---

[11] Spencer (2016).

background demonstrate, educating myself further still might not resolve the issue. I am left with a nonevidentiary choice of whom to trust.

Unless you are a climatologist, meteorologist, or atmospheric physicist, you are in the same boat I am. Belief in anthropogenic global warming hinges mostly on whether one accepts the authority and integrity of the scientific establishment, including the soundness of academic publishing, the impartiality of peer review and funding, and resistance of individual scientists and institutions to confirmation bias. For many people, especially liberals and progressives, science is the only trustworthy institution remaining in our society. To doubt anthropogenic climate change is to question the very source of legitimate truth in our culture; as well, it is to question the other institutions that draw their legitimacy from science.[12] That is why, especially in the United States, those who disbelieve in climate change are generally members of the religious right who also disbelieve in other, even more fundamental, scientific theories. If you already believe that evolutionary theory is a vast unholy conspiracy to deny the biblical creation story, it isn't much of a stretch to disbelieve in climate change as well. There is some truth in the derisive association of climate doubters with flat earth believers.[13] The truth is not in the derision though, because

---

[12] By science here I am not intending to impugn the Scientific Method itself, only whether the institutions of science uphold it faithfully. Whether their failure to faithfully uphold it reflects deeper epistemological and ontological problems is another matter. The Scientific Method carries tacit metaphysical assumptions (such as an observer-independent objective reality) that are untestable from within its assumption-set. The apparent failure to uphold the objective pursuit of truth in its institutions may be an irremediable reflection of the limitations of its metaphysical foundation, rather than a conditional shortcoming that could in principle be eliminated through reforms to peer review, academic practices, more stringent replication of experiments, and so forth.

[13] I know several very intelligent people who believe the earth is flat. The recent popularity of Flat Earth Theory reflects a growing public alienation from the scientific establishment. Most commentators attribute this either to the arrogance and poor communication skills of scientists, to the inaccessibility of highly specialized scientific language, or to the stupidity and ignorance of the public. Another possibility, though, is the scientific establishment has earned this distrust through its alliance with the Establishment generally, whether economic or ideological. P.S. I think Earth is round. P.P.S. To the extent, that is, that the objective "is" of identity is ontologically valid.

what is happening is not that they are silly or stupid. It is that they are rebelling against the dominant culture's primary epistemic authority.

Another factor that might predispose someone to disbelieve climate change is that it might conflict with deeply held economic, social, or political views. Unsurprisingly, most climate change doubters hold conservative political opinions. They typically oppose government regulation of business and see climate change as a dangerous justification for increased regulation. They usually favor unbridled exploitation of "natural resources," deriding the idea that nature poses any limits to human growth that technology cannot overcome. They are usually pro-nuclear power, pro-fracking, pro-offshore drilling, pro-coal mining, and in favor of bringing industrial development to the entire planet. Quite often (though not always), their position that we aren't harming the climate is of a piece with their position that we aren't harming the environment generally; that we shouldn't worry too much about GMOs, chemical waste, nuclear waste, plastic in the oceans, pesticides, pharmaceutical waste, habitat destruction, and so forth. Furthermore, climate-change-doubting blogs and especially their comments sections are often peppered with Islamophobic sentiments (the government is using the climate change hoax to distract us from the real threat: Islam!) and other alt-right canards.

Here, in short, are two nonevidentiary reasons to believe in anthropogenic climate change: faith in the institution of science, and the bad company of those who doubt it's happening.

So what was the final result of my descent into the world of climate skepticism? If you are still waiting for the answer to "Which side am I on?" I'm afraid you'll have to wait a little longer (until the end of this chapter). One thing I found in my excursion, however, is that each side is mistaken in the characterization of the other. The skeptic side, while certainly surrounded by a penumbra of ignorance, pseudoscience, and worse, is home to many reasonable, scientifically literate individuals who endure intense hostility for articulating heterodox viewpoints. The "War on Evil" approach to combating climate skeptics (starting with the poisonous slur "climate denier") is based on false premises. While I think that they sometimes overlook or minimize data that doesn't support their position, prominent dissidents like Judith Curry, John Christy, Roy Spencer, Jim Steele, and Stephen McIntyre are neither corrupt, stupid, nor insincere, and at least some of them are also passionate environmentalists who care deeply about the ongoing degradation of nature. Moreover, at least from the perspective of a layman who has looked at both sides, some of their criticisms have merit. Whether or not the mainstream view is right, science and the public would benefit from a more respectful and less dogmatic engagement with the skeptics.

The skeptics' derisive view of establishment scientists is also wrong. It is obvious to me when I speak with climate scientists and read scientific papers that these people are also, generally speaking, scrupulous, conscientious scientists who care deeply about the planet. When skeptical bloggers accuse them of being part of an evil conspiracy, of criminal negligence, financial corruption, or hidden "political agendas"; when they bandy about degrading caricatures of "greenies" and "enviros," they undermine the credibility of any legitimate criticisms they may have.

Furthermore, many skeptics who are not trained scientists are frequently guilty of intellectual sloppiness of the grossest kind, which suggests that *they* are the ones with a political agenda. They uncritically embrace flimsy evidence and arguments that serve their desired conclusions. To give a representative example, I came across an authoritative-looking graph of ice core proxy temperatures going back thousands of years, apparently from ten thousand years before present to today, showing that temperatures during the Minoan Warming, Roman Warming, and Medieval Warming were much higher than present temperatures.[14] It was presented in a right-wing blog that essentially said, "The climate establishment must be idiotic or corrupt, when their own data shows that present temperatures are far below historical periods." The comments section was

---

[14] I do not want to reproduce the graph, because I'm afraid someone will use it without context. You can easily find it on the internet by searching for "GISP2 Ice Core Temperature Data Last 10,000 Years."

a chorus of agreement. It was an impressive graph, so I went to look at the information source, which was a peer-reviewed paper by R. B. Alley.[15] There I saw that the graph created by the blogger was highly misleading, because the data series from which it drew only went up to 1905 (which would make sense since ice cores are not useful proxies for very recent temperatures). Yet the graph was labeled to look as if it went up to the present day. All it showed, then, is that historical temperatures were much higher than they were in 1905—before modern emissions-caused warming is supposed to have begun.[16]

Of course, the behavior of a cadre of scientifically untrained and politically motivated followers doesn't entail that the skeptics' arguments are without merit. It should caution us to proceed carefully though, and to be aware of confirmation bias—our own as well as that of others. Confirmation bias refers to the tendency to prefer evidence that conforms to an existing belief and to interpret evidence in a way that supports that belief. So, the right-wing bloggers embraced that graph without subjecting it to any scrutiny whatever, even though a cursory check of the underlying data revealed it as bogus.

The more ego attachment one has to one's opinions, the greater the likelihood of confirmation bias. Signs of this ego attachment include self-righteousness, smugness, and contempt for those who disagree. I am sorry to say that I see a lot of all three in the writings of both sides, leading me to have little trust for either. Go read the blogs and comments sections of each side and ask yourself whether these people would be open to being wrong.

Now you, dear reader, may think yourself relatively free of confirmation bias, but notice how you respond when you read something critical of your position on climate change. Don't you subject it to much greater scrutiny than you would something that supports your position? Who is that guy? Was it in a peer-reviewed journal? Is he funded by oil companies? Let

---

[15] Alley (2000).

[16] This sloppiness obscures the fact that the Minoan, Roman, and Medieval Warm Periods were probably warmer than present temperatures.

me find something that debunks it.... From that mindset, it takes only the most superficial rebuttal, character assassination, unsubstantiated accusation, etc., to cause the believer to dismiss the criticism. By the same token, you will probably give a free pass to articles that confirm your position. You won't bother to look at the unadjusted raw data, to question the fidelity of proxy temperatures, and so on. Generalize this tendency, and we have a society of increasingly noncommunicating reality bubbles, warring with each other even as their hidden agreements go unexamined, and their shared interests neglected.

The End of the World

A politically progressive friend described her experience of spending a week with her in-laws, who consumed a steady diet of Fox News. By the end of the week, she said, she understood how it seemed to them that anyone who voted for Hillary Clinton must be an idiot. The conservative media creates its own reality bubble.

The same might be said for the world of climate skepticism, and for its mirror opposite, the world of climate catastrophism. I encourage the reader to spend some time in each of these reality bubbles. Anchored by scientists and writers like Guy McPherson, Paul Ehrlich, Paul Beckwith, David Wallace-Wells, and Malcolm Light, the catastrophist camp criticizes mainstream climate science along many of the same lines as the skeptics do. It says that scientists ignore data that doesn't fit their worldview, or for which they are psychologically unprepared. Even when they do realize that it is already too late, political expediency induces them to tone down their forecasts; privately, they are much more pessimistic than their public statements indicate. IPCC reports are similarly watered down under political pressure. The truth, they say, is that we are doomed.

Oddly enough, climate skeptics and climate catastrophists come to a similar place of inaction from entirely opposite directions. What does it matter, when one party disengages because they think there is no problem, and the other disengages because they think there's no solution?

Apocalyptic thinking in general fosters a complicity with the very systems that it critiques. Seemingly radical, the catastrophist position is in practical terms completely compatible with the continuation of business-as-usual. Making a similar point, the scholar Eileen Crist writes:

> *Indeed fatalism is a mind-set that strengthens the trends that generate it by fostering compliance to those very trends. The compliance that fatalism effects is invisible to the fatalistic thinker, who does not regard him or herself as a conformist, but simply as a realist.*[17]

The "realism" upon which so much climate discussion is based takes for granted many of the same beliefs and systems that are generating the crisis to begin with. What we believe to be real, though, may be a projection of the story we live under. As for the systems, humans created all of them. Humans can change all of them.

Catastrophist prognostications of doom range from massive disruptions that would render the tropics uninhabitable and devastate food supplies, all the way to near-term extinction of human beings (in my lifetime) or even a runaway greenhouse effect that would make Earth like Venus. I invite the reader to browse Guy McPherson's website, "Nature Bats Last," for a catalog of the scientific evidence behind their position. Basically, near-term extinction depends on positive feedback loops that accelerate climate change. For example:

- Arctic warming melts undersea methane hydrates, releasing methane into the atmosphere and causing more warming.

- The same occurs for stores of methane and carbon dioxide in permafrost.

- Hotter temperatures generate more water vapor, which traps more heat.

- Arctic ice melt decreases albedo (reflectivity), generating more warming from the sun.

---

[17] Crist (2007), 54.

- Warming causes shifting climatic patterns, leading to forest fires and peat fires, creating soot that dirties the snow, causing faster melting.

- Methane release from bodies of freshwater increases with higher temperatures.

- More atmospheric CO_2 leads to more carbonic acid in the rain, which dissolves calcium carbonate rocks, releasing more CO_2 into the air.

Most of the alarm centers on methane. According to Malcolm Light, the methane under the Arctic Ocean alone is a hundred times greater than that sufficient to instigate a major extinction event.[18] If even 1 percent is released, it would cause a 10°C rise in global temperature—enough to ensure the demise of all vertebrates.

And, say the catastrophists, this is already well under way and irreversible. The feedbacks are already in place. The Arctic will soon be ice-free. The Larsen B and Larsen C ice shelves are on the verge of collapse. The West Antarctic Ice Sheet is losing 150 cubic kilometers of ice per year. The oceans are warming at twice the rate previously thought. Sea level rise has gone exponential.

I will not repeat the previous exercise and walk the reader through the mainstream responses to these points, the responses to the responses, and so forth. Methane levels haven't risen as quickly as the catastrophists predict. Yes, they have—the methane has gone to a higher atmospheric layer than where the measurements are taken. No, they haven't—that claim is speculation based on sketchy data. Yes, they have...

I seriously recommend the interested reader spend a solid week reading catastrophist literature, and another solid week reading skeptic literature (the website *Watts Up With That?* is a good place to start, or Matt

---

18 Light (2014).

Ridley's essay "The Climate Wars' Damage to Science").[19] It is amazing how intelligent human beings, all sourcing information from what we call science, can come to such dramatically opposed conclusions. What's going on here? Each camp wields various psychological and political theories to explain the intransigence of the other. Each side is certain that the science is with themselves.

For reasons that will become apparent in this book, I do not accept the catastrophist narrative. It does, however, have three important truths to offer.

First, a great dying is indeed under way on this planet, and human activity is responsible for it. Most people and institutions have their heads in the sand and do not see it or allow themselves to feel it.

Second, we are indeed facing the end of the world. Not the literal end of civilization or the human species, but a transition so profound that on the other side of it, it will seem like we are living in a different world. That is how deep the changes must go for the ecological crisis to be resolved. We face an initiation, a metamorphosis, into a new kind of civilization. From this place, what is possible, practical, and realistic changes as well. Our successful graduation to a new world is by no means guaranteed; nonetheless, the catastrophists are channeling the truth of a possibility. They see the necessity of a death phase, the dying of our present collective self; they do not see the rebirth. And that is normal. In a true initiatory ordeal, often there is a moment when there seems no hope of ever making it through.

Third, the catastrophists are right that conventional means, methods, and mindsets are far insufficient to the task of healing the planet. The catastrophists are like the voice that tells the man in the maze, "Just stop." They do not recognize that after this stopping a new compass becomes available, a song that can guide us out. The situation is hopeless, yes—but only from within the logic and worldview that entrap us. That worldview (which has

---

[19] Ridley (2015). Perhaps in an effort to establish its credibility, this essay indicts as pseudoscience other deviations from conventional opinion, such as belief in homeopathy or the dangers of genetically modified food.

generated the crisis to begin with) renders us impotent, because its solution set is entirely insufficient to the task at hand.

Many of my readers have probably had at least one experience in their lives that violated what they'd believed to be possible. A precognitive dream, a healing of an "incurable" disease, an uncannily accurate psychic reading, an amazing synchronicity, an encounter with a UFO—something that implied "reality is much bigger than we've been told." If you are one of them, I ask: Does your despair take that into account? Or do you exclude such considerations from your "realism"?

Ironically, some catastrophists in their despair have indeed hit upon a significant theme of the song that can lead us out. They are saying that since it is hopeless, we might as well dedicate our lives to love, beauty, and life. Yes! That is the starting point, because our current predicament is the result of a long history of denying love, beauty, and life. The revolution is love. What becomes possible then?

Translated into practical action, this change of heart is ultimately more important to healing the climate than the things the conventional alarmists are calling for. It is as if giving up on saving the world opens us up to doing the things that will save the world.

The Institution of Science

If the skeptical "right" and doomsaying "left" are both trapped in reality-tunneling confirmation bias, perhaps we should flee to the center: the standard climate change narrative. This is comfortable territory, staked out by our society's primary epistemic authority, science.

The problem is, the dynamics that afflict the two extremes afflict the middle as well. Over the last few years, a growing chorus of insider critics have been exposing serious flaws in scientific funding, publishing, and research, leading some to go so far as to say, "Science is broken."[20]

The dysfunctions they describe include:

---

[20] Belluz and Hoffman (2015).

- Various kinds of fraud: some deliberate, but mostly unconscious and systemic[21]

- Irreproducibility of results and lack of incentive to attempt replication[22]

- Misuse of statistics, such as "P-hacking"—the mining of research data to extract a post hoc "hypothesis" for publication[23]

- Severe flaws in the system of peer review; for example, its propensity to enforce existing paradigms, to be hostile to anything that challenges the views of the reviewers whose careers are invested in those views[24]

- Difficulty in obtaining funding for unorthodox research hypotheses[25]

- Publication bias that favors positive results over negative results, and suppresses research that won't benefit a researcher's career[26]

The system encourages the endless elaboration of existing theories about which there is consensus, but if one of these is wrong, there are nearly insuperable barriers to its ever being overturned. These go far beyond classic Kuhnian resistance to paradigm shift—critics call it "paradigm protection." Former NIH director and Nobel laureate Harold Varmus describes it this way:

> *The system now favors those who can guarantee results rather than those with potentially path-breaking ideas that, by definition, cannot promise success.*

---

[21] Freedman (2010).

[22] Baker (2016).

[23] *The Economist* (2013).

[24] Smith (2006) and *The New Atlantis* (2006).

[25] McNeil (2014).

[26] Peplow (2014).

Young investigators are discouraged from departing too far from their postdoctoral work, when they should instead be posing new questions and inventing new approaches. Seasoned investigators are inclined to stick to their tried-and-true formulas for success rather than explore new fields.[27]

It is easy to see how these dynamics might impact climate science, a politically charged field that receives billions of dollars of government funding. Skeptics' websites contain laments by climate researchers who are afraid to attempt publication of results that contradict climate orthodoxy because they do not want to be ostracized as a "denier"; professors telling of discouraging graduate students from investigating inconsistencies in the data; and anecdotes about reputable scientists who lost funding and professional appointments after they issued mild criticism of official positions.

The dissident climatologist Judith Curry raises questions about the genesis of the scientific consensus around climate change:

The skewed scientific "consensus" does indeed act to reinforce itself, through a range of professional incentives: ease of publishing results, particularly in high impact journals; success in funding; recognition from peers in terms of awards, promotions, etc.; media attention and publicity for research; appeal of the simplistic narrative that climate science can "save the world"; and a seat at the big policy tables.[28]

All of this adds up to a kind of collective confirmation bias within science, the same cognitive handicap that so obviously afflicts many climate skeptics. In other words, confirmation bias is not limited to those outside the establishment. It is institutionalized within it as well, despite the system of peer review that is supposed to eliminate it. Here is what my father, a retired professor, says about peer review:

Peer reviews in my field were often sloppy, dashed off because reviewers had little incentive to spend time. No one received the authors' data to replicate.

---

[27] Albert et al. (2014).

[28] Curry (2016).

Editors could bias the result through choice of reviewers (this is important). Also, coteries of researchers in specialized fields, who were the only ones who could understand a given article, would make favorable reviews to enhance the status and visibility of their clique.

Let me hasten to add that this doesn't mean the establishment view on climate (or anything else) is wrong. It means, though, that if it were wrong, we may not easily know it. We would know it only if the self-correcting mechanisms of science-as-institution are properly functioning.

To those who suspect me of being "anti-science," let me make a confession. The consensus around global warming that brings together Big Science, governments, and most of the world's elites makes me less confident, and not more, in the standard narrative.

Why should I accept the consensus around climate change when I reject the very same consensus that has been invoked in support of GMOs, nuclear power, pharmaceutical oncology, or the safety of common pesticides?[29]

The reader might object that the consensus on these topics is weaker than the consensus on climate change, and she may be right. The prospect of offering stronger examples of questionable scientific consensus presents a bit of a quandary, however. If I reveal my doubts about, say, standard Big Bang cosmology, dark matter, the Lipid Hypothesis for arteriosclerosis, or the pumps-and-channels model of cell membrane physiology, then I will be undermining the credibility I need to make my point effectively. The reader will assume I am deficient in intellect, ignorant of basic science, or credulously enamored of kooky theories. He will lump me in with biblical creationists, flat-earthers, and moon landing conspiracy theorists. Or perhaps he will conclude my contrarian views have a psychopathological origin; that I'm rebelling against my father or suffering from oppositional defiant disorder.

It is impossible to cite an example of a fallacious scientific consensus that will be convincing to a person who trusts scientific consensus. Of

---

[29] Mitch Daniels's 2017 *Washington Post* op-ed, "Avoiding GMOs Is Not Only Unscientific, It Is Immoral," exemplifies this genre. See also my response: Eisenstein, 2018.

course, one could adduce historical instances where scientific consensus was wrong—the luminiferous ether, eugenics with its calls to save humanity from genetic degradation, and the hackneyed example of geocentric cosmology come to mind—but the believer can turn those around and say, "See, science works. Wrong theories are eventually rejected and we are converging on the truth." The implication is that the big mistakes are all safely in the past.

None of this is to say that I believe in every scientific heresy I encounter. After all, many scientific heresies are themselves mutually contradictory. On many issues I don't have a strong opinion one way or the other, because when I try to pin it down and figure out which side is right, I end up in a welter of competing claims that I am incapable of evaluating—just as I described in the satellite temperature debate.

The reader may be familiar with this kind of rabbit hole. Whether you are investigating 9/11 conspiracy theories, chemtrails, crop circles, vaccine damage, or nonstandard archaeological, cosmological, biological, or geological theories, the pattern is the same. One side invokes the authority of the scientific establishment, while the other consists largely of marginalized heretics. These dissidents complain about the difficulty they have obtaining research funding, getting published in journals, and getting their arguments taken seriously. Meanwhile, the defenders of orthodoxy cite the self-same lack of peer-reviewed journal publication as reason not to take unorthodox theories seriously. Their logic is basically: "These theories are not accepted; therefore they are not acceptable." That is confirmation bias in a nutshell.

In most controversies that pit a powerful orthodoxy against a marginalized heterodoxy, the establishment makes liberal use of scare quotes and derisive epithets like "denier," "conspiracy theorist," or "pseudoscientist" to exercise psychological pressure on the undecided layperson, who does not want to be thought a fool. These tactics invoke in-group/out-group social dynamics, leading one to suspect that the same dynamics might prevail within the scientific establishment to enforce groupthink and discourage dissent. But again, perhaps the unorthodox theories really are bunkum and deserve the derision directed at them. We the laypeople cannot know. It comes down again to our trust in authority.

I would like to advance a narrative of ecological healing that does not depend on trust in existing institutions of authority, scientific or otherwise. Science can still be an ally (I will draw heavily from it in the next two chapters) but it need not be the master.

In the extremely polarized climate debate, it might be hard for the reader to actually believe that I am not attempting to construct a surreptitious case against anthropogenic global warming. That is not my intention. To repeat: my intention is to uncover hidden agreements shared by all parties to the debate, agreements that will generate a worsening crisis and, ultimately, catastrophe no matter which side is right.

The Wrong Debate

I'm sure by now you are impatiently awaiting my opinion as to which side is right, ready perhaps to breathe a sigh of relief when I excuse the foregoing as an intellectual exercise and assure you that of course, I do believe in climate change. Which side am I on? Okay, here is a summary of my opinion, which I will elaborate throughout this book:

We are in fact facing a very serious climate crisis. However, the main threat is not warming per se; it is what we might call "climate derangement." This derangement is caused primarily by the degradation of ecosystems worldwide: the draining of wetlands, the clear-cutting of forests, the tillage and erosion of soil, the decimation of fish, the destruction of habitats for development, the poisoning of air, soil, and water with chemicals, the damming of rivers, the extermination of predators, and so on. Through disruption of the carbon cycle, the water cycle, and more mysterious Gaian processes, these activities degrade the resiliency of the ecosphere, leaving it unable to cope with the additional greenhouse gases emitted through human activity. The result may or may not be continued global warming, but it is certain to bring increasingly wild fluctuations not only in temperature but also, more importantly, in rainfall. (This may already be happening, as evidenced by the recent spate of record hot *and* cold temperatures in various places around the world.)

Standard climate theory gives primacy to CO_2-induced radiative forcing as the cause of climate change, relegating ecosystem degradation to secondary status. In standard climate theory, radiative forcing (the greenhouse effect) warms the atmosphere by only a little over 1 degree Celsius for each doubling of carbon dioxide. That by itself gives little cause for alarm. What is alarming is the potential amplification of this heating through a host of positive feedbacks. I will argue that these depend much more on biological processes than we have realized. When biological systems are degraded, they lose their ability to adapt to changing climate and to maintain stable conditions under which they can thrive.

The problem with the climate debate then, is primarily one of misplaced emphasis. Whether average global temperatures are increasing is not the main issue. We are engaged in the wrong debate. Climate derangement will continue even if we stop emitting carbon, and it will bring calamity even if average temperatures remain constant. That is because Earth is a living body, not a machine, and we have been destroying its tissues and organs.

Anthropogenic climate derangement began long before the industrial era, primarily through deforestation and soil erosion. In the last centuries these have reached industrial scale, while greenhouse gas emissions present a whole new challenge that a seriously degraded biosphere is poorly prepared to meet.

Let me put my thesis starkly:

If the standard narrative of AGW is true, then the most urgent priority is to protect and restore soil, water, and ecosystems worldwide.

If the standard narrative of AGW is false, then the most urgent priority is to protect and restore soil, water, and ecosystems worldwide.

One purpose of this book is to justify these assertions, and to describe the shift in perception and mythology that will support their enactment.

As for the normative climate debate, on the most primal level my sympathies are with the alarmists. Whatever the flaws in their data, arguments, and models, the basic alarm that animates their fervor is well founded. If average temperature stops rising or reverses, we should be no less alarmed. Moreover, the characterization of skeptics as "deniers" also has a core of truth. But it isn't their skepticism about the science that makes them deniers; it is the denial of the ecological holocaust, the decimation of Earth's biological wealth and vitality.

It's like this: Suppose I were infected with a flesh-eating bacteria that is killing me, and everyone is arguing over whether I have a fever or not. Those who say "Yes, he has a dangerous fever. We'd better take care of him" are closer to the truth than those who say "He doesn't have a fever, so he must be fine." Now, my condition might indeed be accompanied by a dangerous fever, and it might make sense to take down the fever. But if the flesh-eating bacteria is not stopped, I will die soon anyway, whether by fever or something else. For the planet, the flesh-eating bacteria is the global financial system, and underneath it the Story of Separation. Development and extraction are devouring the world.

If you are a climate skeptic reading this book, I want you to snap out of your denial. That doesn't mean getting on board with climate science. It means opening your eyes to the ruin of so many precious places, to the wounds of strip mines and oil spills and toxic waste sites, to destruction of habitats and species, to the impoverishment of life on earth. It means feeling the agony of this planet, taking it in, and doing something about it.

In my lifetime the number of monarch butterflies has dropped by 90 percent. Fish biomass has dropped by more than half. Deserts have expanded to an unprecedented extent. Coral reef extent has declined by half. Mangroves in Asia have declined 80 percent. The Borneo rainforest is nearly gone. Rainforests globally cover less than half their former area. Thousands of species have gone extinct. All that is real, and it is just a trace of the degradation afflicting this planet. Be alarmed. We cannot lose many more of the planet's organs and tissues before calamity strikes.

If you are a climate alarmist, I applaud your alarm and ask you to shift its focus. Alarm needn't depend on whether human survival is threatened. To me the prospect of humanity persisting on a dead, denuded planet is more alarming than a future without humans. How would you like to be the sole survivor of a holocaust in which all your friends and family perish? "What will happen to us?" is, I will argue, too small a question, and the kind of alarm that comes from it is too narrow and, in the end, counterproductive.

Whichever side you are on, I'd like you to hear a different alarm. It is about the dying of life on this planet. Have you noticed as I have a marked decrease in windshield bug splatter when driving? When I was a kid, I remember the windshield being covered with bug splatter. I wondered if it was memory that was faulty, until I read a twenty-seven-year study documenting a 78–82 percent decline in flying insect biomass in protected nature reserves.[30] It is a thorough, extensive, and scrupulous study that echoes similar findings around the world.[31]

If I were in charge, this study would be a screaming front-page headline. Insects were the first animals to colonize land, arriving around the same time as plants. They are crucial to every terrestrial food chain. Insects are woven deeply into life. Fewer insects means less life. It means the planet is becoming less alive. Let me rephrase that: It means the planet is dying.

No one knows the cause, but the authors note that it probably isn't warmer temperatures, as these correlated during the study period to more insect biomass not less. They cite chemicals and diminished habitat in nearby farmland as possible causes. I think that is likely, and that a deeper cause lurks underneath. It is that we are not treating the world as alive and sacred.[32] We have not acted in service to life. We have instead seen the rest of life as the servant of man. That is what wants to change. The ecological crisis provides the initiatory medicine for the world's dominant civilization to make that change. The crisis will intensify until the medicine has been fully received.

---

[30] Hallmann et al. (2017).

[31] For an introduction to other research documenting the insect apocalypse, see Hunziker (2018).

[32] By "we" here, I mean the dominant civilization. To the extent that you participate in it, that "we" also includes yourself, even if you dissent from its systems and beliefs with all your heart.

4

The Water Paradigm

A Different Lens

One thing that gets lost in the climate debate is that Earth's climate is already severely deranged. This is hard to see when the conversation is about global averages and the predictions of computer models. But severe climate change is already devastating the lives of millions of people. To see it, we have to look through a different lens—not temperature and carbon, but water.

In recent decades, the word "climate" has increasingly become a proxy for "temperature." Read almost any discussion of the droughts and floods that are striking nearly everywhere on the planet with increasing frequency, and you'll see climate change mentioned as a major—if not *the* major—culprit. Traditionally, though, it was as common to speak of a wet or dry climate as it was a hot or cold climate. Increased levels of drought and flooding are not *caused* by climate change; they *are* climate change.

While most of the discourse around climate change focuses on temperature, water is the climatic factor that most directly impacts life. Life flourishes throughout the hot equatorial zone because of the presence of abundant rainfall, while deserts, because they receive little precipitation, are comparatively barren whatever their temperature.

The ability of land to support human life also depends on water. The more regular and abundant the precipitation, the better able the land is to sustain large numbers of people. A hotter-than-average summer is usually no great threat to crops; a drought threatens catastrophe.

Of course, temperature bears a strong influence on precipitation patterns, most directly through its effect on wind and ocean currents. Moreover, the water cycle and the carbon cycle are closely entwined. We cannot speak of one without speaking of the other. The shift of emphasis I am about to offer is nothing as simplistic as "Water is more directly impactful, so we should forget about carbon." What we will see is that by putting water first, the carbon problem and the warming problem will be solved as well.

Water vapor is the dominant greenhouse gas on the planet, accounting for 80 percent of the greenhouse effect. Its effects are hard to model, however, because unlike carbon dioxide it is not evenly distributed throughout the atmosphere. Furthermore, when it condenses into clouds, water exerts a cooling effect by reflecting sunlight during the day, as well as a warming effect by insulating the surface and absorbing long wave infrared radiation, especially at night, depending on the type and height of the cloud. The evaporation and condensation of water also transfers heat from lower layers of the atmosphere to higher layers, and horizontally from one region to another. The interplay of these regionally variable effects is what makes water difficult to accurately model.

Making it harder still is one critical factor: life. Until recently, rainfall patterns and cloud formation were thought (by scientists) to be primarily a function of geophysical processes. Where there happened to be ample rains, life flourished; where there was little rain, drylands formed. This view is at home with the deeper belief that the planet is a host for life, but is not itself alive; that life is but a fortuitous biological scum atop an inanimate rock.

James Lovelock and Lynn Margulis's Gaia Theory, which posits that life creates the conditions for life, put an end to the conceptual separation between geology and biology. As this paradigm percolates through science, it encourages a new perceptual stance that reveals things that were previously invisible—invisible to scientists, that is, although not to traditional and indigenous people.

The paradigm shift when it comes to climate is not really from carbon to water; it is a shift from a geomechanical view to a Gaian view, a living systems view. Whether we are looking through the lens of carbon or water, from the living systems perspective we see that climate health depends on the health of local ecosystems everywhere.

The health of local ecosystems, in turn, depends on the health of the water cycle, and the health of the water cycle depends on the soil and the forests.

Local ecosystem
↓
water cycle
↓
soil + the forests

The Forests and the Trees

An alive planet is a resilient planet, capable of responding to fluctuations in atmospheric gases, volcanic eruptions, asteroid impacts, solar fluctuations, and other challenges. Standard climate theory says that forests have an ambiguous contribution to temperature, contributing to warming because they absorb more sunlight than bare ground, and to cooling by storing carbon. Recently, the trend in research has been to demonstrate that forests store and sequester much more carbon than previously thought. According to one paper, if we continue current rates of deforestation then the planet will warm by 1.5 degrees even if fossil fuels were eliminated overnight.[1] These computations did not include lost sequestration potential, but only the carbon from the lost biomass and exposed soil. (Deforestation exposes soil to heat and erosion, leading to massive emissions of carbon dioxide.)

On carbon grounds alone, forest conservation and reforestation should be much higher priorities than they are today. Through the lens of water, their importance is even more critical.

---

[1]   Mahowald et al. (2017).

Because forests store and transpire moisture, they convert solar radiation into the "latent heat" of water vapor. Some of this heat is released back at night when the water vapor condenses into dew, but a lot of the vapor rises to form clouds, transferring heat from the ground level into the atmosphere. When the water condenses as clouds, the heat is released again. How much of that heat radiates out into space and how much returns back to earth is a contentious matter—the effect of clouds is one of the most important and controversial variables in climate modeling[2]—but there is little doubt that forest transpiration has a cooling effect on at least a local and regional level; there are also strong arguments that the same is true on a global level.

Intuitively, everybody already knows that it is much cooler in the forest (during the day, and a bit warmer at night). Research confirms this commonsense knowledge. One study in the Czech Republic compared air temperatures under conditions of high solar irradiance (i.e., sunny days) on neighboring parcels of wet meadow, harvested meadow, asphalt, forest, sparse vegetation, and open water. The air temperature over the wet meadow, lake, and forests was cooler than 30 degrees; the harvested meadow was over 40 degrees and the air above the asphalt nearly 50.[3]

These are local effects; forests also apparently cause regional cooling. Kenya, which has lost most of its forest cover over the last half-century, is also suffering persistent droughts and higher temperatures. Regions in Kenya where the daytime temperature in the forest might be 19 degrees record temperatures in nearby, recently cleared agricultural land of 50 degrees.[4] In Amazonia, pastureland was found to be on average 1.5 degrees hotter (day and night combined) than forested areas despite its higher

---

[2] Generally speaking, clouds with lower cloud tops radiate more heat back into space. See Trenberth and Stepaniak (2004).

[3] See Ellison (2017) for an image showing these findings, which originally appeared in Hesslerová et al (2013).

[4] Schwartz (2013).

albedo.[5] In Sumatra, land cleared for palm oil plantations was 10 degrees hotter than nearby rainforest, and stayed hotter even when the palm trees matured.[6]

A real, living forest interacts with the water cycle in complex ways that science is just beginning to understand. One way is by converting humidity to rain. Water vapor in the atmosphere doesn't necessarily fall as rain, but may instead persist as haze in what is known as a "humid drought." One reason for the formation of haze is an overabundance of small condensation nuclei, which prevents water droplets from becoming large enough to fall as rain.[7] Pollutants, smoke from forest fires, and dust from desiccated soil are among the culprits in haze formation. Over forests, the condensation nuclei are mainly biogenic, including plant detritus, bacteria, fungal spores, and secondary organic aerosols originating as volatile organic compounds emitted by vegetation.[8] These aid the formation of clouds rather than haze, and allow cloud formation at higher temperatures than abiotic nuclei.[9] Recent research confirms the increased cloud cover over and near forests.[10] These lower, thicker clouds have a greater cooling effect than high-altitude clouds. According to one researcher, a 1 percent increase in albedo from forest-generated clouds would offset all warming from anthropogenic greenhouse gas emissions.[11]

On the other hand, the haze that forms in the absence of forests exercises a powerful greenhouse effect. It lets in the sunlight and covers the

---

5   Runyan and D'Odorico (2016).

6   Sabajo et al. (2017).

7   Runyan and D'Odorico (2016), 62.

8   Ibid.

9   Thompson (2008).

10   Teuling et al. (2017).

11   Jehne (2007).

earth in an insulating blanket that prevents heat from radiating back into space at night. The result is intense heat and humidity, but no rain. This demonstrates the principle that life creates the conditions for life.

Some of the bacteria that serve as cloud condensation nuclei seem almost custom-designed to seed clouds. The most studied species, *Pseudomonas syringae,* bears ice-nucleating proteins that allow clouds to form at higher temperatures (and thus lower altitudes) than they otherwise could. Found around the world, they originate as plant pathogens.[12] Their ice-nucleating proteins lead to frost damage on plants, which they are more easily able to feed on. Ominously, crop scientists are working to genetically engineer strains of *Pseudomonas syringae* that lack the ice-nucleating proteins. This is a typical control-based approach that may have the utterly unanticipated consequence of altering rainfall patterns and intensifying climate change.

Deforestation sets off a vicious circle of drought, extreme weather, and even more deforestation. Familiarity with the water cycle makes it clear why. In a healthy water cycle, evaporated water from the ocean moves over the continents, where it falls as rain. A tiny fraction of that precipitation runs off directly; most of it is absorbed by soil and vegetation, while some percolates into underground aquifers, eventually surfacing as springs that feed streams and rivers. Once the water is in the soil and water table, plants and especially trees steadily transpire it back into the air, providing a source of rain through the dry season. Depending on the region, some 30–90 percent of rainfall originates not directly from the ocean, but from evapotranspiration of water from soil and vegetation.

In vast areas of the earth, trees are critical to the ability of soil to absorb rainwater:

- The leaf litter layer absorbs water and protects moisture from immediate evaporation.

- The shade canopy also slows evaporation.

---

[12] Schiermeier (2008).

- Trees and forest fauna increase the porosity of the ground, allowing water to penetrate.

- Tree roots and understory vegetation protect soil from erosion.

Deforestation, on the other hand, leads to soil erosion and the reduced capacity of the land to absorb water, and consequently worse flooding after heavy rains. Furthermore, without the deep roots of trees to bring moisture from deep underground and replenish atmospheric moisture, droughts tend to be longer and drier. This in turn puts greater stress on remaining forests, which become susceptible to fires and disease. When the rains do come, they run off the parched earth, taking the soil with them.

Deforestation alters atmospheric circulation in another way: it leads to stronger updrafts and higher clouds, which produce rainfall that is less in total quantity but greater in intensity—aggravating the familiar drought/flooding cycle.[13] The transition from reliable rainfall to the drought/flood pattern exemplifies the aforementioned "climate derangement" that may be a bigger threat than outright global warming. Not only do weather patterns change, but Earth's ability to handle those changes diminishes.

It gets worse. Forests do more than recycle moisture originating in the oceans; apparently, they actually generate wind patterns that bring the water from the oceans in the first place. It has been a commonplace belief around the world that forests bring the rain, but for a long time scientists scoffed at this notion: forests grow where there is ample precipitation, they said, but they do not cause that precipitation. It comes via winds that are governed by large-scale geomechanical processes originating in polar/equatorial temperature differentials, the spin of the planet, and other factors. Now this view is changing.

In the last decade, a scientific theory called the "biotic pump" has been gaining prominence that validates the universal vernacular wisdom that

---

[13]   Schellnhuber (2004), 253.

forests attract rain. First proposed in 2006 by Russian physicists V. G. Gorshkov and A. M. Makarieva, the theory says that evapotranspiration from large forests, especially old growth forests, creates low pressure systems when the water vapor rises and condenses.[14] Because winds generally blow from high pressure to low pressure areas, moisture-laden winds from the ocean are pulled toward forested continental interiors, bringing the rain that in turn maintains the forest.[15] That is why forested continents enjoy reliable, abundant rainfall deep into the interior; that is also why these rains have begun to fail as deforestation approaches critical levels in Amazonia, Southeast Asia, Africa, and Siberia.

The theory sparked intense controversy, as is common when a challenge to long-established dogma comes from outside a discipline (Gorshkov and Makarieva are nuclear physicists, not atmospheric physicists). It is also difficult to prove, either experimentally or through computer modeling; furthermore, it suggests that existing climate models are neglecting an extremely important process. It also carries alarming implications given high levels of deforestation globally. For example, it means that Amazonian deforestation will not lead to a mere 15 percent or 30 percent decrease in rainfall, as conventional models predict, but to as much as a 90 percent decrease.[16] This would mean a transition of Amazonia not to a savanna but to a desert.

Indirect evidence for the biotic pump abounds in the form of droughts and declining rainfall that accompany deforestation from Siberia to Australia to Indonesia to Central America. Total rainfall in the Amazon declined from 1975 to 2003 by an average 0.3 percent a year,[17] in direct correlation

---

[14] The best introduction to the theory and its significance that I've found is this interview with the authors in Hance (2012).

[15] Gorshkov and Makarieva (2006).

[16] Schwartz (2013). ·

[17] Courcoux (2009).

to deforestation rates, culminating in severe droughts in 2005, 2010, and 2015. More recently, direct evidence has accumulated as well, based on precipitation patterns and isotope analysis.[18] While the theory defies the geomechanical bias that still exerts strong influence in climatology, it resonates deeply with the living planet perspective. Again, life creates the conditions for life.

Even in the conventional carbon frame, rainforest conservation should be more prominent for its storage and sequestration of carbon. In the living system frame, to preserve and restore forests is a matter of utmost urgency. Today, the number one priority of conventional environmentalism is emissions reduction, but that is the convenient problem, fitting comfortably into the familiar blueprint narrative of the onward march of technology. But the ecological crisis will not be resolved by adjusting a few inputs. We are called to deep partnership with nature and respect for all life.

Crucial forests are tipping into a death spiral: deforesting causing drought, drought causing more deforestation. We have to start protecting forests as if they were sacred (they are), and restoring damaged forests as if our lives depended on it (they do).

The connection between forests, water, and life has always been obvious to people living in deep connection to the land. Here is the Yanomami shaman Davi Kopenawa describing the destruction of the hydrological cycle:

> We never tear away the earth's skin. We only cultivate its surface, because that is where the richness is found. In doing so, we follow our ancestors' ways. The trees' leaves and flowers never stop falling and accumulating on the ground in the forest. This is what gives it its smell and its value of growth. But this scent disappears quickly once the ground dries up and makes the streams disappear into its depths. It is so. As soon as you cut down tall trees such as the wari mahi kapok trees and the hawari hi Brazil nut trees, the forest's soil becomes hard and hot. It is these big trees that make the rainwater come and keep it in the ground.... The trees that the white people plant, the mango trees, the coconut

---

[18] See, for example, Angelini et al. (2011) and Andrich and Imberger (2013).

trees, the orange trees, and the cashew trees, they do not know how to call the rain.[19]

Note the last sentence, which asserts that a forest is more than a collection of trees. If we do not see forests as living beings too, will we ever treat them as such?

The necessity of conserving and re-growing forests is undeniable when we see Earth as a living being and the forests as one of her vital organs. The necessity of protecting and revering the water is obvious when we see it as the blood or vital fluid of a living planet. It is the same as for the human body: If you understand it as a coherent, intelligent living system, then you do not need physiological reasons to convince you that yes, you do need your lungs, your liver, your appendix, your tonsils. It is only in a mechanistic view that we would imagine that some organs are useless and could be cut out without repercussions for the whole. Finally today more enlightened doctors are realizing this and overturning seventy years of medical fads like the routine removal of appendixes, tonsils, and wisdom teeth. Isn't it time we do the same for the Gaian body?

The Organs of Gaia

Forests are certainly not the only Gaian organ crucial to the maintenance of life. Based on the principle that life creates conditions for life, the most important organs would be the ones that are most abundant with life: forests, wetlands, estuaries, coral reefs, and rich grasslands with their vast herds of animals. All are in steep decline around the world, while the biota-poor areas—the deserts and oceanic dead zones—are spreading.

The carbon fundamentalist paradigm has brought welcome attention to wetlands, forests, seagrass, and prairies, which have enormous carbon storage and sequestration capacity. The ten-foot-thick topsoil of the American Midwest testifies to that capacity and to the disastrous consequences of

---

[19] Kopenawa and Albert (2013).

tilling the land and exposing the soil to erosion and oxidation of its organic matter into CO_2. I will look at these non-forest ecosystems, together with cultivated land, through the carbon lens in the next chapter.

Looking past carbon to water and beyond, we see even more clearly the acute planetary importance of these ecosystems. Virgin grasslands exercise many of the same functions that forests do, effectively soaking up rainfall and protecting soil, preventing flooding, ameliorating drought, seeding cloud formation, and building the water table. The thick mat of sod softens the impact of rain on soil, preventing erosion; the carbonaceous soil organic matter that the roots deposit over time is a sponge for rainwater, binding it to organic molecules and slowing evaporation as well.

Just as a forest is more than a collection of trees, a grassland is more than a concentration of grass. It is a living ecosystem that also includes herbivores, predators, and invertebrate multitudes. Earthworms aerate the soil and produce rain-storing humic aggregates; herd animals crop, trample, and fertilize tall grasses that then become mulch and eventually soil. Fungi bind earthworms, bacteria, roots, insects, and each other into complex communities that cycle nutrients and exchange chemical information. Each member of the grassland is alive, and the totality is alive too.

If forests, grasslands, wetlands, coral reefs, etc., are among Gaia's vital organs, then perhaps species can be seen as her cells and tissues. They may not have a visible, direct effect on carbon or water cycles—but then again they may. A traditional Navajo proverb went, "Without the prairie dogs, there will be no one to cry for rain." That seems like bald superstition, except that the near-extirpation of prairie dogs in the twentieth century indeed coincided with declining rainfall in the American Southwest. And now it turns out that the Navajo belief was not so superstitious after all, but rather an astute insight into ecological hydrology. Bill Mollison, preceptor of the permaculture movement, wrote, "Amused scientists, knowing there was no conceivable relationship between prairie dogs and rain, recommended the extermination of all burrowing animals in some desert areas planted to rangelands in the 1950s 'in order to protect the sparse desert grasses.' Today the area has become a virtual wasteland."[20] Mollison offered the explanation that the burrows of prairie dogs and other animals are like lung alveoli. As the moon passes overhead, tidal forces bring up water from aquifers closer to the surface, providing moisture for rain. Judith Schwartz adds that prairie dog tunnels allow rainwater to penetrate the ground instead of running off, thereby replenishing aquifers;[21] prairie dogs also control water-hogging mesquite.

Wetlands, as the name suggests, are also crucial for a healthy water cycle. They slow the migration of water from land to sea, giving it time to soak down into aquifers and evaporate up into the atmosphere to be a source of rainfall. Wetlands have been in decline throughout history as human beings drained them for agricultural purposes, a process still under way today. The current landscape of North America, with its brooks, streams, and rivers running through well-defined channels, is actually the result of severe land modification. According to researcher Steve Apfelbaum, "Many currently identified first, second, and third-order streams were identified as vegetated swales, wetlands, wet prairies, and swamps in the original land survey records of the U.S. General Land Office."[22] Owing to civil engineering projects (such as the straightening of meandering rivers for navigation) as well as the near-extirpation of beavers, the slow progress of water from land to sea was greatly hastened: flow rates of rivers increased by orders of magnitude. Globally, this means that the land is losing water faster than it receives it, making drought an inevitability and contributing to the rise of sea levels.

Ironically, a large part of wetlands destruction in recent times is done in the name of fighting climate change, since big hydroelectric projects often involve severe hydrological disruption. The African Sahel was once home to vast, fecund wetlands of incredible biodiversity, fed by seasonal floods.

---

[20] Quoted in Buhner (2002).

[21] Schwartz (2016), 82.

[22] Apfelbaum (1993).

They have been in steep decline since the dam-building era began in the 1980s, encouraged by development agencies to generate electricity and control floods. As a result, Lake Chad stands at 5 percent its former surface area. Social disruption has followed, fueling Boko Haram and waves of migration to Europe. Next in line is the Inner Niger Delta, a vast wetland the size of Belgium threatened by a new mega-dam planned in Guinea.[23] Writing in *Yale Environment 360,* Fred Pearce observes, "Dried-up wetlands are often blamed on climate change when the real cause often is more human interference in river flows."[24] How convenient it is to blame climate change, compared to questioning a basic strategy of Third World development.

I will mention two more biomes here that are normally excluded from that category: agricultural land and urban land. As I will discuss later, the healing of this planet is not a matter of humanity stepping out, creating a separate human realm and leaving nature untouched. It will not come through minimizing our impact; it will come through changing the nature of our impact. It will come through a different kind of participation in nature, one where humanity returns to being an extension of, and not an exception to, ecology.

As it stands, wherever modernity has spread, lands heavily influenced by humans are damaged lands, ailing lands, unable to fulfill their function in maintaining Gaian homeostasis. Naked soil, such as that plowed up for farming, is almost never seen in nature, and for good reason. It is like an open wound, flesh without skin, that quickly loses its life-giving moisture and blows away. Baked by the sun and lacking a root structure to hold it and aerate it, it can neither absorb as much moisture when it rains, nor hold that moisture very long thereafter. Chemical-intensive agriculture adds injury to injury by destroying earthworms and other soil organisms that help water penetrate to deeper soil layers. Not only do earthworms

---

[23] Pearce (2017).

[24] Ibid.

increase soil moisture capacity, they and the soil ecosystems they facilitate increase soil carbon storage and promote the growth of methanotrophs— bacteria that feed off methane and reduce levels of this greenhouse gas.[25]

Not only do naked, disturbed soils hemorrhage carbon into the atmosphere, they also contribute to direct regional warming: one study notes the correlation between the increase in cover cropping in the Canadian grain belt, lower summer temperatures, and higher humidity and rainfall.[26] Cover cropping is part of a growing regenerative agriculture movement that seeks to restore water and soil through farming.

Other modern agricultural practices that compound the damage to water and soil include:

- The creation of large, unbroken fields without hedgerows, wild or wooded patches, or contours to slow down water and stop erosion during heavy downpours

- The use of heavy tractors, which compact the soil and make it less permeable

- Irrigation that makes soil increasingly salty

- Heavy use of chemical fertilizers, herbicides, fungicides, and insecticides that destroy soil biota

These and other unsustainable practices will stop when we understand that human well-being is inseparable from the well-being of soil and water.

In urban environments the damage to soil is even more severe; often it is paved over entirely. Unable to filter into the earth, the water is a nuisance that is sluiced away through drainage systems as "wastewater," returning swiftly to the ocean without entering the hydrological cycle of evapotranspiration or groundwater recharge. Meanwhile, cities draw down surrounding water resources to meet their use needs.

Without much vegetation to transpire water and cool the air, cities are subject to the urban heat island effect. The heat affects wind patterns,

---

[25] Biodiversity for a Livable Climate (2017).

[26] Ibid.

generating high pressure systems that push precipitation into surrounding areas—for example, cooler mountain regions—which then experience torrential downpours, erosion, and flooding.[27] To a lesser degree, any devegetated area (such as plowed fields) becomes a heat island that generates high pressure and pushes rain away to mountains or the ocean.

Climate change skeptics sometimes cite the heat island effect to claim that global temperature data is skewed, since temperature gauges are increasingly located in or near urban heat islands. Even if true, that is of scant consolation if the whole planet is becoming a heat island due to urbanization, development, and deforestation. The effects are not only local; through their disruption of hydrological heat transport they influence global temperature as well as drought and flood, often through complex, nonlinear chains of causation. For example, deforestation and wetlands draining along Europe's Mediterranean coast have led to decreased evapotranspiration and fewer summer storms near the coast but more intense storms in central Europe. Fewer coastal storms then leads to salinization of the Mediterranean and changes in the Mediterranean-Atlantic salinity valve, which in turn intensifies Atlantic storms and changes weather patterns as far away as the Gulf of Mexico.[28]

As the paramount environmental narrative today, climate change obscures the much larger, more direct, and more local influence of "land management changes" in causing drought, flooding, heat waves, and other kinds of extreme weather. Climate change, instead of being an incentive to enact more ecologically beneficial policies, becomes a convenient scapegoat that diverts attention from effective local measures and shifts responsibility for ecological healing onto distant, global institutions.

For example, understanding that deforestation and soil tillage lead to topsoil erosion, which makes the land unable to absorb rainwater, which then leads to flooding, one necessarily must respond locally: conserving

---

[27] Kravčík et al. (2007).

[28] Millán (2014).

forests and wetlands, practicing organic no-till agriculture, and rebuilding soil. Ignorant of these things, the environmentally concerned person is left with actions like putting solar panels on the roof or offsetting jet travel by donating to a tree planting fund. Environmental zeal stays focused far from home, and most of the damaging activities continue.

I write these words in the aftermath of Hurricane Irma and Hurricane Harvey, proclaimed by the media to have been exacerbated by climate change. While I understand the scientific logic behind that claim—warmer water evaporates more quickly, warmer air can hold more water, etc.—the argument appears tenuous under close examination.[29] Total accumulated cyclone energy has not appreciably increased in recent decades, nor has total precipitation, storm frequency, or storm strength. Regardless, the controversy over whether climate change is responsible deflects attention from local factors that make such storms more damaging to humans and ecosystems. Chief among them, at least in Florida and Texas, is the widespread draining of wetlands, which can soak up rainwater and buffer storm surges. Both regions have also experienced deforestation, agricultural soil abuse, and extensive suburban development. Blaming climate change obscures these factors and allows these practices to proceed as usual.

As with flooding, so with drought. I recently read an otherwise insightful article on immigration by Vijay Prashad that stated, "The causes [of emigration from Central America] should be found in the collapse of agriculture in these countries—driven largely by climate change induced drought and flash floods, extreme heat and forest fires."[30] Let's leave aside for a moment the economic and political causes of the collapse of agriculture, such as free trade agreements that make traditional peasant agriculture uneconomic, benefit transnational agribusiness, and turn the agricultural economy toward exportable commodities. While global climatic patterns (namely, the strong El Niño of 2015–16) precipitated

---

29   See NOAA (2018).

30   Prashad (2017).

the latest famine, these countries have also suffered intense deforestation. Guatemala lost 17 percent of its rainforest in just fifteen years from 1990 to 2005; subsequently the rate of deforestation accelerated threefold;[31] losses have been especially heavy in its famous cloud forests.[32] A similar story transpired in Honduras, which lost 37 percent of its rainforests in the same period with no letup in sight. El Salvador is the saddest case of all, having suffered 85 percent deforestation since the 1960s. When these forests are cut down, rainfall runs off instead of being absorbed to recharge the water table, resulting in erosion, landslides, and flooding. Springs dry up, rainfall decreases, and the local climate becomes hotter and drier. The stage is set for devastating drought.

Before deforestation, the rainforests of South and Central America received plenty of rainfall, El Niño or not. That's why they are called rainforests. Moreover, El Niño (a weather pattern that brings drought and heat waves to much of the northern hemisphere) has been rising in frequency and intensity since the 1970s. Typically blamed on "climate change," it may also be a by-product of deforestation, particularly in Indonesia, where severe deforestation may weaken the persistent zone of low pressure that helps drive the Walker circulation, whose weakening results in El Niños.[33]

Blaming climate change for Central American droughts diminishes the urgency of addressing local deforestation, shifting emphasis onto global-scale solutions. It relegates to secondary importance a whole complex of other causes far beyond deforestation. Besides, what causes deforestation? Whether in Central America or elsewhere, the causes can include:

- Changing weather patterns caused by previous deforestation and soil degradation

---

[31] All of these figures come from Mongabay (2018), which has a heartbreaking catalog of deforestation trends around the world.

[32] Community Cloud Forest Conservation (2018).

[33] Hance (2012).

- International free trade agreements, which make traditional, sustainable farming practices economically unviable and demand conversion of forests into ranches and monocrop plantations

- The ideology of development, which makes traditional, sustainable peasant agriculture seem backward

- The erosion of indigenous land-based spiritualities that held protecting land and water as a sacred duty

- Foreign debt owed by forested less-developed countries that pressures them to convert those forests into commodities

- Establishment of legalized property rights in places where informal communitarian land ownership was an obstacle to development

- The extermination of large predators that kept herbivores in population equilibrium with their environment

- Government policies that bring nomadic and indigenous people into the mainstream of industrial society, so that they can no longer tend the wild

- Population pressure leading to the cutting of trees for firewood

- Illegal logging facilitated by "corruption"—which is actually the incursion of translocal monetized relationships into preexisting gift-based structures

- Unpredictable knock-on effects of ecological disruption caused by draining wetlands, spraying chemical poisons for "weed" and "pest" control, and exterminating crucial species like beavers, prairie dogs, and wolves; elephants, rhinos, and lions

Obviously, these are not isolated dysfunctions in a fundamentally sound system. The system itself, and the Story of Separation woven through it, generates the dysfunctions. If pressed to distill them down to a single culprit, I would say it is the truncating, the simplifying, and the impoverishing of relationships—human to human and human to world. And if I were pressed to offer a universal solution, it would be to see and treat the world as sacred again.

If anything on earth is sacred, it should be water. So far I have actually not upheld it as sacred in this discussion; I have merely illuminated the bad things that are happening to ourselves and the planet through the maltreatment of water, trees, and soil. To treat them as sacred we must go beyond that. As my friend Orland Bishop says, the sacred is something that requires a sacrifice; that is, it is something we value—and would sacrifice to protect—beyond its use-value to ourselves.

Other cultures upheld the sacredness of water through ceremonies and taboos, protecting water from anything that would insult or pollute it. I do not advocate imitating indigenous ceremonies; rather, we can find a contemporary counterpart that draws from their knowledge and fits into our own evolving story-of-the-world. Our water technologies will take on the energy of ceremony when they draw from the perception that indigenous and traditional people have held of water, that it is a living being. The door to that is opening as the familiar scientific conception of water as a homogeneous, structureless chemical fluid becomes obsolete.[34]

The hydrological arguments of this chapter offer a nudge toward treating water as sacred, but do not touch on other water issues that are more difficult (to our current way of knowing) to connect to climate change. One day, though, I am sure that we will learn that contaminating water with pesticides, pharmaceutical residues, industrial chemicals, and radioactive waste will threaten planetary well-being just as much as deforestation or greenhouse gas emissions. Water is life. What we do to water, we do to life.

Five Thousand Years of Climate Change

The Standard Narrative of climate change contends that climate was relatively stable until the twentieth century, when industrial emissions started to become significant. Whether or not this is the case in terms of temperature,

---

[34] See my essay "The Waters of Heterodoxy" (Eisenstein, 2014) for an in-depth discussion of this point.

in terms of water the last few thousand years have seen dramatic changes in climate. The earth has become significantly drier, and I am afraid that yes, much of the blame rests on the shoulders of human civilization.

According to some researchers, the buildup of CO_2 and methane was well under way long before the Industrial Revolution. William Ruddiman claims that anomalous (compared to previous interglacials) buildup of both gases coincided in its onset with Neolithic deforestation and land cultivation.[35] His paper assembles diverse evidence—historical, archaeological, and geological—that massive deforestation had occurred by two thousand years ago in China, India, the Middle East, Europe, North Africa, and to a limited extent in the Americas. Its contribution to greenhouse gases, he says, is double that of the industrial era, which has merely accelerated a long-term trend.

Ruddiman adopts a conventional greenhouse lens in discussing the issue; from the perspective of water and the biotic pump, the situation is even more alarming. Have you ever looked at a satellite image of the globe and felt a chill of foreboding seeing the vast and growing deserts stretching eight thousand miles from the west coast of Africa through the Arabian Peninsula all the way to Mongolia? Plus their smaller cousins in the American Southwest, the west coast of South America, and most of the continent of Australia? Not to mention southern Africa, and now even parts of Spain and Brazil? Most of these places were once green. Mongolia became desert just some four thousand years ago, not millions of years ago as previously thought.[36] The Sahara was a lush savanna six thousand years ago. Geomechanically oriented scientists typically attribute its desertification to a shift in the tilt of Earth's axis, but human activities probably exacerbated it.[37] As recently as Roman times, elegant cities stood in what is now desert, nourished by long-gone forested

---

[35] Ruddiman (2003).

[36] Yang et al. (2015). Read more at Yirka (2015).

[37] See, for example, Weisman (2008).

watersheds.[38] The Middle East, the cradle of civilization, was likewise once a fertile paradise; deforestation there is recorded as far back as the *Epic of Gilgamesh,* as well as in pollen and charcoal deposits. Biblical forests like the Wood of Ziph and the Forest of Bethel are now desert; gone as well are the cedars of Lebanon and the forests of the Greek isles where Artemis hunted. Deforestation accelerated in Roman times, and is often blamed for the demise of the Roman Empire.

In *Critias,* Plato offers a vivid and accurate description of the effects of deforestation:

> *Now that all the richer, softer soil has been washed away, only the bare ground is left, like the bones of a diseased body. In former times ... the plains were full of soil and there was abundant timber in the mountains.... The annual rainfall used to make the land fruitful, for the water did not flow off the bare earth to the sea.... Where once there were springs, now only the shrines remain.*

In many places the deserts continue to spread, and new deserts to form. Earth lost more than 3 percent of its remaining forests from 2000 to 2012. At present, Earth has only about half of the trees it had at the dawn of civilization.[39] The United States has lost an area of forest the size of Maine in the last decade. Deforestation in Brazil increased by 29 percent in 2016, before dropping a little in 2017 to a rate still higher than 2012. Queensland, Australia, lost a million acres of trees in 2015–16, contributing to the sediment-induced stress of the adjacent Great Barrier Reef.[40] Globally, tree cover loss rose by 51 percent in 2016.[41]

You get the picture: In developed and less-developed countries alike, the extent and quality of forests are deteriorating. Owing as well to other kinds of land and water abuse, desertification is claiming 12 million hectares

---

[38] Hughes (2014), 3.

[39] Crowther et al. (2015).

[40] Robertson (2017).

[41] Weisse and Goldman (2017).

of land per year globally, according to the U.N. Moreover, desertification is just the most conspicuous manifestation of the general impoverishment of life on earth that extends to every region and biome. Life is on the decline almost everywhere, even in places that look nothing like deserts.

In other words, the land is dying before our eyes, as it has been doing since ancient times. We have to stop killing it. This is bigger than cutting greenhouse gas emissions. It is reversing a relationship to soil and sea that has been part of civilization for thousands of years. I am sorry, but merely switching to so-called renewable energy sources is not enough. We are called to visit deep questions like "What are we here for?" "What is humanity's right role on earth?" "What does the earth want?"

As we explore these questions, some of the measures advocated by climate activists will take on new motivation and significance, while others will be revealed as yet another iteration of the old relationship. Big hydroelectric plants, endless landscapes of solar arrays and wind turbines, and in particular biofuel plantations damage the ecosystems they occupy. In the new relationship (new for civilization, though not for the indigenous), whenever we take from the earth, we seek to do so in a way that enriches the earth. We aren't unconscious of our impact, nor do we seek to minimize our impact. We seek to make a beautiful impact that serves all life.

The answer to the above questions (What are we here for?...) that I will explore in later chapters starts with the understanding of this chapter and the next, that life creates the conditions for life. And who are we humans? We are life too. We are life, born into a certain form, with a unique array of gifts. Like all life, our purpose is to serve life—to serve both what it is and what it might become. For never is life static. Each unfoldment of complexity builds on the last. What is the dream of life? What wants to be born next, and how can we serve that? These are the questions that need to replace civilization's former question: How can we most effectively extract resources from the earth to build the human world?

5

Carbon: The Ecosystems View

Carbon, Soil, and Life

In this book I have intentionally placed the water chapter in front of the carbon chapter, signaling the possibility that we need to reverse priorities in our consideration of these two substances so foundational to life. Nonetheless, carbon also offers an important window onto the health of the Gaian organism. Ultimately, it opens onto the same understanding that water does: that we need to shift our attention to ecosystems, soil, and biodiversity.

Most discourse about greenhouse gases focuses on emissions from fossil fuels, how to replace them with alternative energy sources, and whether it is possible to do this quickly enough. This is well-trodden territory. I will not weigh in on it, because I don't want to give more attention to what I consider to be the wrong conversation. Whether or not we reduce emissions, in the absence of ecological healing on every level, climate derangement will continue to worsen.

To enter the ecosystems paradigm through the carbon lens, we must look at an item of the carbon budget that is much less certain than fossil fuel emissions—the release of carbon from "land use changes" (a euphemism for ecosystem destruction)—and, equally uncertain, the capacity of intact ecosystems to absorb and sequester carbon.

Some researchers believe we have drastically underestimated both,[1] and the general trend in the literature has been toward higher and higher estimates for terrestrial fluxes of CO_2. For example, a recent study concludes that tropical deforestation of 2.27 million square kilometers has contributed about 50 gigatons of carbon to the atmosphere since 1950, and the rate of emission has accelerated.[2] According to this study, emissions from tropical deforestation currently total 2.3 gigatons/year—more than 20 percent of anthropogenic emissions, and much more than previous estimates.[3] A similar trend applies to other biomes.

Quite possibly, the effect of deforestation, soil loss, biodiversity loss, swamp and peat bog draining, draining of mangroves, and other land use changes is so severe that one could reasonably argue—even in the carbon (as opposed to water) frame—that climate change is caused by these activities as much as by burning fossil fuels. Fossil fuel burning intensifies the instability that ecological devastation would cause anyway.

I recently read on a climate skepticism blog the claim that from 1750 to 1875, atmospheric CO_2 rose far faster than cumulative anthropogenic emissions, and that the latter didn't catch up with the former until 1960.[4] The author argues that CO_2 rose as a result of rising temperatures (rather than causing them), a common skeptic position. There is another possibility, though: the period in question was also a time of massive deforestation

---

[1]   See, for example, Arneth et al. (2017).

[2]   Rosa et al. (2016).

[3]   Estimates of carbon sunk into rainforest biomass have been increasing. A 2012 paper in *Nature Climate Change* (Baccini et al., 2012) put the figure at 228.7 gigatons, fully 21 percent higher than the 2010 estimate of the *Global Forest Resources Assessment* (Food and Agriculture Organization of the United Nations, 2010). Nonetheless, the *Nature Climate Change* paper's estimate of annual emissions from tropical deforestation is less than half the *Current Biology* figure above, probably because it does not consider below-ground biomass and legacy emissions.

[4]   Middleton (2012).

in Europe and North America, along with the vast expansion of farmland. These sources of CO_2 may have dwarfed fossil fuel emissions.

The previous chapter described the process by which deforestation, conventional agriculture, and other land abuse emit carbon into the atmosphere from exposed and eroded soil. Here are a few examples of the other side of the equation: the capacity of intact ecosystems to absorb and store carbon underground.

Wetlands

Where should I begin the sad news? That the planet has lost half its mangrove swamps in the last century, and about 70 percent of its total wetlands?[5] That seagrass meadows are declining by 7 percent per year?[6] That total wetlands loss in the United States stands at 50 percent since its founding, and has accelerated in the twenty-first century over the twentieth?[7] Most of the losses are due to agricultural conversion, urban growth, and coastal development; meanwhile, intact wetlands suffer degradation from pollution and saltwater encroachment. Rising sea levels ordinarily wouldn't be a problem, because coastal wetlands would migrate, but today levees constrain their spread, even as dams deprive them of sediment to grow.

From the standpoint of biodiversity wetlands degradation is catastrophic, but what about carbon? Wetlands deposit more carbon in soil than any other ecosystem—in the case of seagrass, as high as 20 tons per hectare per year.[8] Together with mangroves and salt marshes, wetlands

---

5   Davidson (2014).

6   Waycott et al. (2009).

7   Fears (2013).

8   Duarte et al. (2013). This figure combines measurements and modeling over a fifty-year timespan for restored seagrass meadows (natural ones would presumably outperform them). It observes that sequestration rates increase exponentially with time (until reaching a limit), implying that many previous studies conducted on short-term replanting projects grossly underestimate seagrass's carbon sequestration potential. Approximately 30–60 million hectares of seagrass worldwide sequestering 20 tons/hectare give as much as a gigaton of carbon absorbed every year—one-tenth as much as anthropogenic emissions.

account for half of biological carbon capture globally according to some estimates.[9] Peatlands are another massive carbon sink, containing as much carbon in their soil as all living biomass on Earth, carbon that can enter the atmosphere when the peatlands are drained, deforested, or burned.

Sequestration rates for these and other ecosystems are typically determined by measuring soil accumulation rates. This method partakes in the basic scientific strategy of isolating variables, rendering invisible the synergistic connections among systems. Mangroves, for example, trap sediment that would otherwise disturb the coral reefs beyond them, possibly making them more vulnerable to bleaching. Seagrass buffers the acidity of the surrounding water, allowing faster shellfish growth. Both shellfish and coral reefs biosequester carbon in their own right. The basic methodology of carbon accountancy—divide the earth into biomes and regions and add up the carbon sequestration of each—inherently underestimates the value of each.

Grasslands

Intact grasslands inhabited by herds of large herbivores have a tremendous capacity to sequester carbon in the soil. Data in terms of tons of carbon per hectare is hard to pinpoint because estimates obtained through measurement and modeling vary by several orders of magnitude, depending on geological conditions, rainfall, types of grasses and whether they are cut, and the presence or absence of herd animals (whether wild or domesticated livestock). Furthermore, sequestered carbon can remain in the soil for varying lengths of time. Some carbon-containing soil organic matter decomposes in a year or two, much of it remains in the soil for a few decades, and some isn't recycled back into the atmosphere for thousands of years (if ever). The ten-foot-thick topsoil of the American Midwest (much eroded today) testifies to the ability of grasslands to store carbon underground.

---

9   Nellemann et al. (2009). Note that the sequestration figures given in this report are lower than more recent estimates.

The highest carbon storage comes from native grass mixes occupied by large, roaming herds of herbivores. Sadly, 97 percent of the original North American highgrass prairie has been converted to cropland, suburbs, and sown pasture. Originally covering 70 million hectares, its carbon-regulating capacity was enormous. Judging by the data, albeit sparse, coming from management-intensive grazing practices that seek to replicate natural herbivore grazing behavior, it is conceivable that highgrass prairie could sequester 8–20 tons of carbon per hectare per year. Today, instead, most of this land is a carbon emitter, because it is cultivated for crops.[10] Plow-based cultivation that exposes bare soil to the air, water, and wind makes its organic matter (carbon) available for oxidation. A similar story has transpired in the steppes of Asia, the veldts of Africa, the pampas of South America, and so on. According to the FAO, up to a third of global grassland has already been degraded.[11] What could be a carbon sink is becoming an emissions source.

Forests

Of all ecosystems, the general public recognizes forests to be crucial for maintaining climate health. At present they absorb about 40 percent of global anthropogenic emissions—and emit at least a third of that due to deforestation.[12] The more CO_2 in the air, the more they absorb—up to a limit. It is as if they are valiantly doing their best to keep the atmosphere in balance. We humans are not helping. According to some estimates, the total number of trees on earth has fallen by nearly half since the dawn of civilization;[13] today, hundreds of thousands of square kilometers of forest

---

[10] For more on this, see chapter 8 on regenerative agriculture. Most estimates are an order of magnitude less than these figures, but on the other hand the American highgrass prairie represents only 2 percent of global grassland.

[11] Food and Agriculture Organization of the United Nations (2009).

[12] Pan et al. (2011).

[13] Crowther et al. (2015).

disappear every year. The losses may be even worse than commonly estimated, as deforestation statistics fail to include fine-level tree loss, which
according to some researchers accounts for two-thirds of biomass loss
from tropical forests.[14] As with wetlands and grasslands, destruction of
forests turns land from a carbon sink to a carbon source.

Less recognized as a problem than deforestation is forest degradation,
primarily a result of logging, insect damage, and forest fires—three factors
that are intimately related. Contrary to timber industry claims, logging renders forests more susceptible to catastrophic fires, not less.[15] As described
in the last chapter, it creates drier conditions by reducing transpiration and
increasing runoff and erosion. Logging also disrupts the ecological balance
that keeps insects in check. Typically, it homogenizes the forest, leaving it
more susceptible to insects and disease, the stumps of felled trees offering
a breeding ground for both. Meanwhile, since logging eliminates standing
dead trees and the hollows of older trees, important habitat disappears,
increasing the risk of insect and pathogen outbreaks. Roads used for heavy
machinery result in soil compaction and the fragmentation of ecosystems,
further reducing resiliency.[16] None of these effects is accurately encompassed by carbon metrics or climate models.

Furthermore, if we understand that forests themselves are living beings
(rather than mere aggregates of living beings), other kinds of damage
become visible. Mycelial networks of phenomenal complexity tie together
all the trees and other plants in a forest, providing a communication network through which trees share information, alert each other to pests,
and sometimes even share resources. Roads chop up this living web into
smaller, disconnected pieces. Conventional logging also prevents trees
from reaching a very great age and from falling and slowly rotting over
a period of decades or centuries. What if the most elder of the trees, the

---

[14] Baccini et al. (2017).

[15] Wuerthner (2016).

[16] See, for example, Sierra Forest Legacy (2012).

grandmother trees, contain wisdom (or, if you prefer, chemically encoded information) that is useful for the forest to endure unusual once-in-a-century conditions? What if the rotting trees host slow-developing fungi that play an important role in maintaining ecological balance? All of these phenomena are much harder to quantify than metric tons of biomass.

In his 2016 book *The Hidden Life of Trees,* forester Peter Wohlleben makes a strong case for forest sentience and the social nature of trees. His team used radionuclide-marked sugars to establish that healthy trees feed sick trees, and that parent trees nourish their offspring. Sometimes the tree community even keeps the stumps of felled trees alive for centuries. They communicate through airborne chemicals as well as through mycelial networks; they also learn, individually and collectively, from experiences with drought and other threats.[17] Some trees form friendships with other trees, cooperating rather than competing for sunlight. Trees also cooperate to create microclimates: in one study that Wohlleben mentions, naturally growing forests maintained temperatures 3 degrees cooler than in managed areas.

Perhaps the ecological crisis that we frame in terms of climate change and global boundaries will be resolved only when it has brought us to a place where we recognize the aliveness of forests and all things. Only then will we have the knowledge and skill necessary to properly care for the tissues and organs of the Gaian body. But aliveness becomes invisible when a forest or other being is reduced to a dataset.

The living being we call a forest includes not just trees, but all the beings that live there. How would one put a number on the contributions of, say, a population of wolves? Apex predators are crucial in maintaining robust ecosystems, even though they make no direct contribution to carbon sequestration. Their contribution is indirect, systemic, and diffuse. In North American forests, the extermination of wolves and cougars has led to burgeoning deer populations, which consume understory vegetation and new saplings, leaving soil exposed, increasing runoff and erosion,

---

17 For an introduction to Wohlleben's work, I recommend the excellent interview by Schiffman (2015).

reducing water retention, and contributing to reduced rainfall during dry seasons and flooding during wet seasons. Changes in ground and understory vegetation also reverberate through insect, fungal, and bacterial communities, making trees vulnerable to insects and disease, and then to fire. Logging, acid rain, ozone pollution, and changing climatic patterns exacerbate these effects in a wicked synergy. For reasons that are unique to each place, forests are in decline all over the world.

I could quote more figures for aboveground and belowground carbon sequestration for different forest types: tropical, temperate, and boreal, primary closed-canopy forests and second growth forests, and so on. But come on, people, do we really need these numbers to know that we must conserve and treasure our precious forests? Even if we could live on a planet barren of them, would we want to? When is the tree-killing going to stop? I am hesitant to go into more numbers, wary of implying that it is the numbers we should be talking about. Will it help to pile on more quantitative reasons for why we should do what we already know we should do? I don't think so.

If we don't yet know the forests are sacred and precious, more numbers aren't going to help.

A forest is a living being of inconceivable complexity. When we reduce it to a small set of generic relationships and numerical quantities, we set the stage for violence: their physical reduction by chainsaws and bulldozers follows their conceptual reduction into measurable quantities and services. That is why I hesitate to frame the value of forests in terms of carbon. It leaves out their noncarbon ecosystem services, as well as their inherent worth, turning the conversation toward numbers.

Reducing a forest to numbers like biomass and sequestration rates is not so different from reducing them to board feet and dollars. It is the same way of thinking. I refuse to proceed with it any longer.

The Emissions Obsession

I have gone into the carbon accounting frame in order to establish that local, intimate, participatory care "works" even within greenhouse gas logic.

In so doing, however, I have dabbled in a dangerous reductionism, subsuming a dizzying array of complex ecological interactions under a single metric: units of carbon. In doing so, I risk implying, "What matters about these ecosystems is the amount of carbon they sequester," and affirming carbon as a valid proxy for ecological well-being. Thus I participate in the ubiquitous equation of "green" or "sustainable" with low carbon.

The crucial role of living systems in maintaining climate stability presents us with good news and bad news. The good news is that our world can survive, that it can potentially adapt to higher levels of greenhouse gases. The bad news is that the ecosystems that can do this are in steep decline all over the world. That means, given positive feedback loops that are already releasing large amounts of carbon and methane from nonhuman sources, climate instability will continue to worsen even if we cut fossil fuel use to zero, unless we also heal and protect the forests, mangroves, seagrass, and so on.

One challenge that climate change skeptics sometimes pose is that CO_2 levels and temperatures were much higher in past epochs than today, and the planet did just fine. The standard rejoinder is that never before have CO_2 levels risen this abruptly. Whether or not that is true, I think it overlooks a more important issue: where did the biosphere's historical resiliency come from? It came from healthy living ecosystems. Life creates the conditions for life, and the modern era has been an era of unprecedented death.

The climate skeptics also state an important, albeit partial, truth when they claim that increased atmospheric CO_2 will result in more plant growth and higher CO_2 uptake. Carbon uptake has indeed been faster than expected, increasing in ten years from 40 percent to 50 percent of fossil fuel emissions.[18] We might be fine—if it weren't for the fact that some quarter to a third of the planet's land surface has been severely devegetated, and most of the rest compromised by human activity. Deserts, monocrop agriculture, and pavement don't sequester much carbon.

---

[18] Carrington (2016).

Ecosystem disruption has turned many areas from carbon sinks into carbon sources. Likewise, the skeptics are right that CO_2 levels were several times higher millions of years ago, and that Earth's climate undergoes natural fluctuations. Tragically, though, habitat degradation, pollution, development, mining, wetlands draining, overfishing, extermination of predators, and so forth have created conditions where plants and the rest of life are no longer able to fully exercise their capacity to maintain planetary dynamic resiliency. Gaia has the capacity to self-regulate—a capacity we are destroying.

Given growing awareness of the crucial role of forests and other ecosystems in regulating climate, as well as the potential that I will discuss later of regenerative agriculture to sequester massive amounts of carbon very quickly (and, more importantly in my view, restore the water cycle), why is the policy conversation so focused on emissions?

Here are a few reasons:

First and most simply, it is a lot easier to measure or estimate fossil fuel emissions than it is emissions from land use changes. While measurements of biomass have improved with new technology and the accumulation of research, they still vary widely from place to place and study to study. In our present political culture, policymakers, treaty negotiators, and regulators need quantitative measures to define goals, agreements, and rules based on a "carbon budget." Emissions therefore fit much more easily into the political culture.

Biosequestration is even harder to measure than biomass. I asked a researcher, Oswald Schmitz of Yale University, why there is so little hard data on carbon sequestration. His explanation was simple: it is much easier to measure and calculate aboveground carbon storage than it is to measure it in soil. This illustrates a general principle: when we rely on metrics to make policy, it becomes biased toward the things that we choose to measure, that we are able to measure, and that are intrinsically amenable to measurement. Furthermore, what gets ignored often corresponds to cultural blindspots and prevailing social, material, and economic practices.

Generally speaking, it is very difficult to measure the total carbon, let alone the sequestration rate, of soil. Most assays include only the first 30 centimeters or 1 meter, yet deep-rooted grasses and other vegetation can store carbon much more deeply than that.[19] Then there is the matter of the molecular composition of the soil organic matter, which determines how long it stays stored in the ground before reentering the carbon cycle. This time period also depends on local conditions, the microclimate, the composition of soil biota, and so forth.

Biological carbon fluxes (and those of other greenhouse gas) do not lend themselves easily to climate modeling compared to emissions. The more climate is understood to interact significantly with life, the more difficult it is to model. Fluid dynamics, the flow of heat and air and currents, is relatively easy to simulate on a computer. That is not true of the processes of life: the way virgin forests maintain microclimates better than tree farms, the role of cloud-seeding bacteria, the effect of earthworms on populations of soil methanotrophs, the effect of whales on ocean nutrient mixing and hence plankton biomass. Such things are hard to model, or indeed even to understand, without decades of close study. We tend to focus on what is amenable to our most familiar tools.

The focus on emissions rests comfortably in the predominant geomechanical view of the planet, which sees Earth as a complicated machine rather than a living organism.

Modern reductionistic methods are well adapted to dealing with complicated (as opposed to complex) systems. In a complicated system like an automobile or computer, though there may be many variables, the variables are more or less independent. If the system isn't working, you can troubleshoot it by isolating and testing the variables one by one. You can also generate predictable macroscopic effects through controlling one or a limited number of variables. Complicated systems are therefore amenable to a piecemeal approach to problem-solving. The whole is equal to the sum

---

[19] Biodiversity for a Livable Climate (2017).

of the parts, and causal relationships are generally linear. To understand and manage a large, complicated system, you divide it up into lots of pieces and assign a team to work on each piece. The entire structure of academia reflects this approach, with its division into relatively autonomous disciplines and subdisciplines.

The top-down control-based approach that works for complicated systems fails miserably to manage complex systems. In a complex system, variables are dependent, causal relationships are nonlinear, and a small change to one element of the system can dramatically alter the whole. No part can be understood in isolation, but only with reference to an extended web of relationships to other parts. In complex systems, the whole is greater than the sum of the parts; therefore any reductionistic analysis of the system will fail to understand it, and attempts to isolate and alter variables will generate unintended and unpredictable consequences.

Bodies, ecosystems, genomes, societies, and the planet are complex systems. It is tempting to view them otherwise—as extremely complicated machines—because then we can apply our familiar methods of top-down problem-solving and feel like we are in control of the situation. The epitome of this illusion is war thinking, as described earlier in this book, and it extends to every technology of control, from border walls to antibiotic drugs to concrete waterways. Each ends up generating awful unintended consequences, usually including the very opposite of what it was attempting to control (immigration, disease, flooding).

Any narrative, like the Standard Narrative of climate change, is a lens that illuminates some things and obscures others. It obscures, unfortunately, some of the very things we need to pay attention to most if planet Earth is going to heal. In the geomechanical view, such things as topsoil erosion, pesticides, aquifer depletion, biodiversity loss, conservation of whales or elephants, toxic and radioactive waste, and so on were once seen (and in many cases still are seen) as relatively inconsequential to climate change. Such oversights are understandable if we see Earth as a fantastically complicated machine. If we see Earth as alive, then we know that of course destroying its living tissue will render it unable to deal with fluctuations of atmospheric components.

shift in priorities

The point here is not that emissions don't matter. It is a call for a shift in priorities. On the policy level, we need to shift toward protecting and healing ecosystems on every level, especially the local. On a cultural level, we need to reintegrate human life with the rest of life, and bring ecological principles to bear on social healing. On the level of strategy and thought, we need to shift the narrative toward life, love, place, and participation. Even if we abandoned the emissions narrative, if we do these things emissions will surely fall as well. *local ecosystems*

The Geoengineering Delusion

From the standard carbon perspective, the world faces a grim future. The necessary drastic cuts in emissions are impossible to accomplish in time to avoid catastrophe. Many climate scientists conclude that the only practical solution is what is known as "geoengineering"—the artificial altering of the atmospheric composition and surface reflectivity of the planet in order to reduce global temperatures. The three most-researched technologies involve dumping large amounts of iron oxide into the oceans in order to absorb and store carbon dioxide, spraying sulfate aerosols into the atmosphere to increase the planet's albedo (reflectivity), and installing millions of carbon-sucking machines to remove CO_2 from the air.

While many scientists (particularly those with an orientation toward systems theory) are highly skeptical of these proposals, mainstream organizations like the U.S. National Research Council have endorsed their development. Possibly, some of these technologies are already being secretly tested through what some people refer to as "chemtrails." I am skeptical of many of the theories advanced by chemtrail researchers, particularly those involving deliberate attempts to sicken the population, but aerial spraying programs are entirely plausible from the geoengineering mindset of climate modification and weather control.[20]

Geoengineering has also attracted widespread mainstream criticism for possible unintended consequences such as ozone depletion, ocean acidification, and reduced rainfall in the tropics. Ecologists are especially concerned. Considering the havoc wreaked by the introduction of a single new species like rabbits in Australia, imagine what the nonlinear knock-on effects might be of wholesale changes to oceanic or atmospheric chemistry. Such geoengineering proposals are attractive only from the perspective of the engineer managing a machine.

I'm especially troubled by sulfate aerosol spraying, which would essentially bleach the sky to a lighter shade of blue. Once we start bleaching the sky to reduce temperature, we won't easily be able to stop, because stopping would induce a very sudden temperature rise. Unless the spraying coincided with strong measures to reduce greenhouse gases (or, in my view, restore ecosystems), it would have to continue indefinitely.

This concern exemplifies a more general problem with geoengineering. If the living planet view I've described is correct, then these cooling measures will allow the real problem to continue unabated. Thinking we have solved the problem, we will be able to proceed with ecosystem destruction, covering up the symptom while worsening the disease. Taking for granted carbon as a proxy for biospheric health, we will be less able to hear the cries of the earth over the hum of the carbon-sucking machines.

Hey, I've got a good idea. With machines to suck carbon and algae pools to make oxygen, maybe someday we can dispense with nature entirely. Maybe someday we can replace every natural and wild thing with artificial substitutes. Hydroponic solutions can replace soil, water filtration machines can replace wetlands, vat-grown meat can replace livestock. We

---

[20] I won't go into the debate here—if you are interested you can watch the documentary *Overcast*. I am agnostic on the issue, although I have seen some odd things, such as a jet leaving an intermittent contrail in mechanically regular two-second bursts across the entire sky. Maybe it was going through alternating zones of different humidity, but the precise regularity of the dashed line it made across the sky was suspicious.

can fine-tune greenhouse gas levels to create just the right temperature. The conquest of nature will be complete.

What scares me the most is not that this is a vain fantasy doomed to failure. What scares me is that we will succeed.

There is a second category of geoengineering that uses life, rather than chemicals, as its instrument. It is a step toward the understanding that life creates the conditions for life, yet mechanistic and reductionistic mindsets still encroach in its application.

For example, the growing understanding of forest carbon storage has birthed schemes for massive, rapid reforestation through drone planting of trees. The numbers look good: more trees means less carbon dioxide.

We must remember though that a forest is more than a concentration of trees. While planting by drone allows a rate of reforestation ten or a hundred times that of human tree planting, it is necessarily less sensitive to unique local conditions. These conditions may to an extent be encompassed by data on soil, microclimate, and so forth, but that data will leave a lot out. Only people who have been in long, closely observant relationship to the land—ideally a multigenerational relationship—can possibly know exactly what to plant to grow a living forest. Without this knowledge, reforestation efforts often fail or exacerbate the problems they were meant to solve. The best known case is in China, where the "Great Green Wall" of trees planted to stop desertification seemed to work initially, as the trees tapped into deep soil moisture. Eventually, in some places these water-thirsty trees used up available water and died. Before that happened, their leaf canopy shut out light to the original grasses and other vegetation that grew there, exposing soil to the very erosion they were planted to prevent.[21]

The lesson here is that what works in one place may not work somewhere else. Top-down solutions are necessarily based on simplifying assumptions and standardized, scalable measures. We must turn away from an attitude of nature-as-engineering-object to one of humble partnership.

---

[21] Luoma (2012).

Whereas geoengineering is a global solution that feeds the logic of centralization and the economics of globalism, regeneration of soil and forests is fundamentally local: forest by forest, farm by farm. There are no standard solutions, because the requirements of the land are unique to each place. Unsurprisingly, they are typically more labor-intensive than conventional practices, because they require a direct, intimate relationship to the land. Ultimately, no geoengineering scheme that fails to bring us back to intimate relationship is going to work. Tree planting should be a first step in tree caring, tree partnership, tree relationship. This would require millions of people, and not just a fleet of drones, to enter the field of forest care. Is that such a bad thing?

Wait, do I hear a voice saying, "Yes, yes, in the long run we need to do these things, but right now we have to act globally and act fast. The slow work of healing ecosystems is not enough, because without immediate global action to dramatically cut emissions, we will pass the tipping point into irreversible self-reinforcing climate catastrophe." As always, the long term is sacrificed to the short term, and the responses that fit most easily into the status quo are prioritized over those that do not. When we call for quick action, we further empower those already in power, who have the wherewithal to act swiftly and globally. And because the crises are unrelenting and the solutions are superficial, the long term never comes. We just end up giving more power to the same political elites, centralized bureaucracies, and general political system that is inextricably wedded to the current regime of ecocide.

The real problem with the above objection though is that the climate benefits of restoration and regeneration are not slow at all, a fact that will become even more clear when I discuss regenerative agriculture. Remember, I have proclaimed myself an alarmist. The time is *now* to align civilization with ecology. But how do we do that, without in our haste falling into patterns of response that intensify the problem?

Ultimately, climate change is challenging us to rethink our long-standing posture of separation from nature, in which we think we can endlessly engineer our way out of the damage we have caused. It is calling us back to our biophilia, our love of nature and of life, our desire to care for all beings whether or not they make the greenhouse gas numbers go up or down. Geoengineering, beyond its catastrophic risks, is an attempt to avoid that call, to extend the mindset of domination and control to new extremes, and to prolong an economy of overconsumption a few years longer. Eileen Crist puts it this way:

> *Even if they work exactly as hoped, geoengineering solutions are far more similar to anthropogenic climate change than they are a counterforce to it: their implementation constitutes an experiment with the biosphere underpinned by technological arrogance, unwillingness to question or limit consumer society, and a sense of entitlement to transmogrifying the planet that boggles the mind. It is indeed these elements of techno-arrogance, unwillingness to advocate radical change, and unlimited entitlement, together with the profound erosion of awe toward the planet that evolved life (and birthed us), that constitute the apocalypse underway—if that is the word of choice, though the words humanization, colonization, or occupation of the biosphere are far more descriptively accurate.[22]*

From the mindset of the Technological Program (to perfect human control of nature), a more detailed understanding of cloud formation and rainfall leads immediately to the desire to engineer cloud formation and control precipitation. Surely we can improve on the random processes of nature! That mindset says, "When we finally understand the workings of nature on a fine enough scale, we will be able to effectively control it." To manage a complicated (as opposed to a complex) system, the first priority is to understand each of the variables. When we've quantified all the causal relationships among them, then we will know how to control the results. This is akin to the belief that when we understand the minute cellular and genetic processes that constitute human physiology, we will be able to conquer disease with specifically targeted medicines. More generally, the holy grail of science has been complete control of reality through its complete reductionistic understanding built atop a "theory of everything."

---

[22] Crist (2007).

Whether in the area of ecology or human health, we are now learning the hard way that better reductionistic understanding of a complex system does not necessarily bring about better control of that system. Understanding of the molecular machinery of cellular function has (judging by the volume of scientific papers) increased exponentially over the past half-century, yet it has borne nothing by way of the long-promised "targeted" medicines to cure cancer and the new wave of autoimmune diseases. The basic (conventional, that is) treatments are much the same as they were in the 1970s, using blunt chemical, surgical, or radiological force to kill the cancer or override the immune system. A parallel in the environmental realm has been the failure of targeted extermination efforts to repel invasive species. Nor have precision "smart bombs" and "surgical strikes" made war any more effective in achieving its stated goals.

Now I am not saying that there is never a time to surgically remove a cancer, wipe out an infection with antibiotics, or suppress an invasive species. There is, in life, sometimes the occasion for a fight. Fighting is not the problem; the problem is the *habit* of fighting, motivated by an inaccurate worldview that rushes to find an enemy to blame. More generally, the problem is not top-down control-based responses; the problem is to default to these responses due to incomprehension of complex, living systems. We then become lost in a maze of unintended consequences, twisting and turning from one emergency to another, haplessly creating the next emergency with our response to the previous. Each solution worsens the crisis it is supposed to solve.

What, then, are we supposed to do? It is fine to exhort environmentalists and policymakers to adopt systems thinking, but frankly, whole-systems thinking is something that we (collectively) really don't know how to do. Our systems are not geared for it, nor our habits of thought, nor our social, financial, and epistemological infrastructure. Our primary modes of problem-solving and knowledge production are fundamentally incompatible with healthy participation in complex living systems. This presents nothing less than a civilizational crisis. That is why climate change is a rite of passage for humanity, an initiatory ordeal. The failure of our customary ways of applying power to address the ecological crisis will graduate us into a new and ancient way of engaging the world. This new and ancient way is already visible in the margins of civilization—in indigenous and peasant cultures, and in much of what we call alternative or holistic. What it lacks, besides a sufficiently broad culture of practice, is a unifying narrative.

The Cult of Quantity

Carbon accountancy has a certain appeal to my inner math geek. I would love to approach the problem of climate change rationally, with familiar cognitive tools. I would divide Earth into various biomes and use the research to estimate the average carbon sequestration level per hectare for each of them. Multiplying by the total number of hectares, I would add up the contribution of each biome to determine a global carbon budget. I would calculate how much of various greenhouse gases we could continue releasing without exceeding the biosphere's absorptive capacity, and therefore how quickly we need to cut fossil fuels, how many trees we need to plant, how many kelp forests, and so on. Each policy option would come with a number. Adding capacity here, subtracting it there, offsetting a strip mine in one place with a new forest somewhere else, we could make rational, climate-friendly decisions.

I hope the image of the smart guy shuffling numbers around the globe as if playing a board game is at least a little unsettling. As the foregoing descriptions of forests show, nature does not consist of a multitude of discrete, independent bits. When we chop nature up into bits in an attempt to understand it, we lose sight of the relationships among those bits. But ecological healing is all about the healing of relationships.

The arithmetic implied by carbon sequestration figures suggests that we can abstract natural and social processes from their context. When we compartmentalize and reduce and denominate natural processes in terms of carbon, what gets left out are the relationships among them. We are in the habit of calling various processes "factors" that contribute to climate change or climate health, but the very concept of a factor is already a problem. One multiplies factors together to obtain a product; change one factor, and the product changes predictably as well. Factors are how we reduce a number to a bunch of smaller, simpler numbers. But complex systems do not resolve easily into independent factors that we can address piecemeal. Our very approach to problem-solving becomes an obstacle to solving the problem.

It isn't only forests whose living complexity far exceeds our ability to measure, quantify, and reduce to data. What number should we give the climate contribution of sea otters? They don't sequester carbon—but they keep down populations of sea urchins, which when unchecked destroy kelp forests that do absorb carbon and alkalize seawater, allowing shellfish to absorb even more carbon.

What number would we give to the climate contribution of coastal fish? The decimation of commercial coastal fisheries has generated population explosions of snails and crabs, which then devastate salt marshes, which sequester carbon. Some biologists suggest that the decline of seaweed and seagrass can be traced to hundreds of years of overfishing, which predates current stressors of coastal ecosystems.[23] Collapsed trophic relationships leave ecosystems more vulnerable to eutrophication and other disturbances. One result is a shift in dominance from seagrasses and long-lived seaweeds to oxygen-choking algae blooms and, consequently, less carbon sequestration. Fish also help buffer acidity by excreting large amounts of calcium carbonate in their feces. According to the 2015 *Living Blue Planet Report,* the ocean fish population has declined by half since the 1970s.[24] Other research puts the decline in total fish biomass at two-thirds over the last century.[25] Increased acidity weakens corals and shellfish, sending ripples of disturbance through the ocean ecosystem. The health of one impacts the health of all.

---

[23] Schellnhuber (2004), 259.

[24] World Wildlife Federation (2015), 7.

[25] Christensen et al. (2014).

And what about the whales? I remember when I was a child during the heyday of the environmental movement, when environmentalism transcended political identity and the phrase "Save the whales" was not yet a derisive cliché. Today, saving the whales is one of those environmental issues that the climate change crusade has marginalized. It seems a sentimental afterthought in the context of stopping climate catastrophe.

When I began researching this topic, I knew that whales must be crucial for the well-being of the planet, an intuition for which I could offer no evidence. How could saving the whales possibly affect greenhouse gases or any other global parameter?

It turns out that my intuition was well founded. First, an important means of removing carbon from the air is through marine ecosystems, the most productive of which are where cold, nutrient-rich deep water wells up to the surface. The nutrients allow kelp and plankton to grow, supporting an entire food chain and cycling carbon into the deep ocean. Today, there are fewer and fewer sites of upwelling deep water, and as a result, growing expanses of "marine desert" nearly devoid of life. This is ordinarily attributed to warmer surface water; an alternative hypothesis (or at least a contributing factor) might involve the decimation of whale populations, which are at a small fraction of pre-whaling levels.[26] Whales bring nutrients up from beneath the thermocline and deposit them near the surface in their feces. That might explain why krill populations have fallen in Antarctic waters where the whales that feed on the krill have been decimated. One would think the krill would flourish once their main predators were gone, but it is the opposite.

Whales also transport nutrients laterally. Many whales, especially blue whales, grow fat in the polar regions before swimming to the tropics to give birth and nurse their young. Conditioned by a geomechanical paradigm, one would think that biological nutrient transport would be insignificant in the vastness of the oceans, but careful research suggests otherwise. I especially recommend the paper "Global Nutrient Transport

---

[26] See, for example, Roman and Palumbi (2003).

in a World of Giants,"[27] which documents the role of megafauna in distrib-
uting nitrogen, phosphorus, and other nutrients, both within the ocean
and onto land. (This is significant in the carbon frame because phosphorus
and nitrogen are key limiting parameters on the biotic uptake of carbon.)
The paper notes that 150 mammalian megafauna have gone extinct since
the Pleistocene, resulting in precipitous drops in nutrient transport across
continents and oceans. Whale populations have declined by up to 99 percent for some species (such as blue whales), and lateral diffusion capacity
for nutrients has declined by 98 percent in the Southern Ocean, 90 percent
in the North Pacific, and 86 percent in the North Atlantic. Without the
whales, it is no wonder that the ocean deserts are spreading. The scenario
on land is equally grim, also due to megafauna depletion: lateral nutrient
diffusion has fallen by at least 95 percent in every continent besides Africa.

Furthermore, whales and other ocean organisms merely by swimming
around contribute enormous amounts of kinetic energy to the oceans,
which according to some estimates equals the contribution of winds and
tides to the mixing of ocean layers.[28] Not only does this bring nutrients to
the surface, but it might contribute to the cooling of the upper layers. The
steep decline of the whales, and more recently the fish due to industrial
overfishing, could easily contribute to warming of the surface layers of the
ocean even if the oceans' total heat content remained the same.

Another avenue by which whales might affect greenhouse gas levels
is through ecological trophic cascades—chain reactions that reverberate
through an ecosystem. According to one hypothesis, the massive decline
in whale populations following the post–World War II whaling boom,
which drove many species to the brink of extinction, deprived orcas of
their main food source.[29] So they turned to smaller prey, including seals,
sea lions, and otters. They so decimated the otters that the otters' prey, sea

---

[27] Doughty et al. (2016).

[28] Dewar et al. (2006).

[29] Whitfield (2003).

urchins, exploded in population and destroyed kelp forests. Kelp forests in turn are important in sequestering carbon and mitigating ocean acidity. Excessive acidity prevents shellfish and coral from growing, removing another carbon sink.

I hope these examples illuminate the impossibility of encompassing the biosphere within quantitative models, and the impossibility of healing the biosphere with policies derived from those models. I hope they also make it clear that the climate is not something separate from the biosphere in which plants and animals live; rather, the climate is an aspect of the biosphere, much more intricately connected to life than science had thought. Therefore, if we are to inhabit a climate suitable for life, we must serve the flourishing of life in all its forms.

Carbon accountancy contributes to the supposition that we can maintain a healthy biosphere by analyzing the carbon contribution of each of its constituents, possibly sacrificing those that have little contribution and cultivating those that have a lot. This mindset makes no sense from a living planet approach. The carbon accountant cannot measure how whale, forest, or wetlands conservation will impact atmospheric CO_2, much less subtler variables like endocrine-disrupting chemicals in the water or microwave radiation in the air. Thus these receive scant attention at the climate policy table.

In order to value what measurements and models leave out, we need another basis upon which to make our choices. It is the living planet paradigm, based on the Story of Interbeing that says that the health of all depends on the health of each.

As the research cited in this chapter shows, quantitative science can serve the living planet paradigm, elucidating the connections among all beings. However, we can no longer hope to steer policy by quantifying the greenhouse impact of one choice versus another. Such calculations are fraught with peril, because they depend on the accuracy and extent of our knowledge. Species or systems once thought to have little relevance to climate turn out to be crucial. It was only in the 1990s that the role of mycorrhizae in carbon sequestration began to be appreciated. It was only in 2009 that researchers confirmed that fish excrete calcium carbonate.[30] It was only in the 2000s that it became known that boreal forests promote low-altitude cloud formation.[31] The megafauna nutrient transport paper was published in 2016. If we'd accounted for these things using the science of a decade ago, their carbon credit score would have been zero.

Until recently, both boreal and mid-latitude forests in climate models were shown to contribute more to warming than to cooling.[32] Imagine the effects of a "science-based" policy that drew from those models. Imagine you are a timber or mining company executive trying to justify the destruction of a forest for profit. Maybe you have a conscience. You really want logging to be okay, and here is a reason.

Indeed, the lumber industry is fond of saying that forestry products are helping fight climate change, because the trees that are cut down end up as building materials that don't return carbon to the atmosphere for hundreds of years; meanwhile, the trees that grow to replace them take additional CO_2 out of the atmosphere. If that's all you measure—the embodied carbon in wood—the argument is sound. But it leaves out the hard-to-measure soil carbon, erosion, hydrological effects, biodiversity effects, and so on.

Our models and metrics will inevitably be faulty, sometimes underestimating and sometimes overestimating the climate effects of various natural and human activities. In a metrics-based policy environment, developers and polluters will seize on these discrepancies whenever they coincide with their economic interests. Wielding the numbers, their arguments will glow with the sheen of science.

While it might seem to us that, from a climate view, a mangrove is more important than a temperate forest, or a coral reef more important

[30] Kwok (2009).

[31] Spracklen et al. (2008).

[32] Bonan (2008).

than a mountainside, or a wolf more important than a bluebird, we have to remember the limits to our knowledge. Carbon arithmetic will always leave something out.

What else have we left out? I suspect that the scientific lens has excluded far more than this chapter has suggested. Whales' contribution to nutrient cycling is already hard to measure—what about the role of their songs in maintaining a pan-oceanic neural network? What about the role of elephant migration patterns in maintaining the subtle energetic pathways of Earth's ley lines? Surely it would disqualify me from serious conversations on climate if I were to take this opportunity to bring in the cetacean mind, dolphin psychic communication, and the teachings of animal whisperers, plant spirit dreamers, dowsers, shamans, etc., who purport to bring messages of relevance to planetary healing. To the numbers guy, these seem a fatuous distraction from the practical matters at hand. But the knowledge and the methods of the numbers guy are failing us, and another worldview is knocking at the door. I believe that ultimately, planetary healing will require engaging sources of knowledge and ways of knowing far beyond what we accept today.

Even considering only mundane ecological science, the last two chapters have made it clear that unless we bring the excluded, devalued, and marginalized back in, climate change will continue to worsen even if we succeed in drastically reducing fossil fuel use. It might be warming, it might be cooling, it might be intensifying fluctuations, a derangement of normal, life-supporting rhythms. Then will we realize the importance of those things that we'd relegated to low priority: the mangrove swamps, the deep aquifers, the sacred sites, the biodiversity hotspots, the virgin forests, the elephants, the whales—all the beings that, in mysterious ways invisible to our numbers, maintain the balance of our living planet. Then perhaps we will realize that as we do to any part of nature, so, inescapably, we do to ourselves.

6

A Bargain with the Devil

Hazards of the Global Warming Narrative

I am afraid that, in adopting climate as their keystone narrative, environmentalists have made a bargain with the Devil. At first, climate change seemed a boon to environmentalism, a potent new argument for things we have always wanted, a new reason to shut down the strip mines, to conserve the forests, and ultimately to end the expansion of a consumerist society. Finally we had a do-or-die reason to implement agricultural practices that regenerate the soil; restore forests and wetlands; build smaller homes in higher density communities; implement economies of reuse, upcycling, and gift; foster bicycle culture; and spread home gardening. Accordingly, environmentalists welcomed the climate narrative as a useful ally, a legitimizer of things they wished people would embrace on their own merits.

We the environmentalists thought, "What we've wanted to do, now they'll *have* to do." The premises of the conversation shifted away from love of nature and toward fear for our survival. We moved from the heart to the mind, asking that we be motivated by distant consequences—sea levels by the year 2050—rather than by damage that stares us in the face (the fish are gone, the eels are gone, the trees are gone, the whales are gone). Furthermore, we must accept the actuality of these consequences on the word of the scientific establishment. This, at a time when many people (like those in Flint, Michigan, who were told by officials wearing the mantle of science that their water was safe, and like the millions who believe that medical science has failed them) feel betrayed by science and by authority generally.

Let me review some other reasons why I think environmentalists have made a bargain with the Devil:

By resorting to climate change arguments to oppose fracking, mountaintop removal, and tar sands excavation, we put ourselves in a vulnerable position should global warming come into doubt. We might face not monotonic warming but increasingly unstable gyrations that are impossible to attribute convincingly to a single cause. What happens if Earth enters a cooling phase? Does that mean we should put environmental protection on pause? Certainly not, but that is the implication when global warming is the keystone environmental issue. As the resiliency of the skeptics demonstrates, climate change is difficult to prove. While reasonable people can doubt the reality of global warming, there is no doubt that continued degradation of ecosystems will damage and eventually destroy Earth's ability to maintain climate homeostasis. By invoking climate change as the reason to implement conservation policies, we substitute a hard-to-prove reason for an easy-to-prove reason.

The climate narrative globalizes the issue of "the environment," demoting local environmental issues to secondary status. If the reason for saving a forest is the CO_2, then one could rationalize its destruction by promising to plant another forest somewhere else. In a global framing, faraway people can make the changes. Not me. Not us.

If advocates of fracking or nuclear power can argue plausibly that their technology will reduce greenhouse gas emissions, then by our own logic we must support those too. This has already happened: the "Think about it" campaign touted the climate change benefits of natural gas. Hillary Clinton spoke in praise of "clean coal." Advocates of genetically engineered crops promise new organisms that will increase carbon sequestration. Giant hydroelectric projects continue to devastate communities and ecosystems around the world. Perhaps worst of all, vast tracts of land in South America, Africa, and Asia are being bought up by corporations with the intention of converting them to biofuels production to produce supposedly carbon-neutral energy. None of these practices can withstand careful scrutiny; nonetheless, they seem plausible enough to bestow upon their subjects a pro-environmental gloss.

By focusing on temperature and CO_2, we encourage potentially disastrous geoengineering schemes like dumping iron oxide into the oceans or sulfuric acid into the atmosphere. We imply that a technical tweak to CO_2 levels or albedo will solve the problem without a fundamental change in our relationship to the planet, and promote the idea that we can endlessly engineer our way out of the consequences of our actions.

The argument that climate change is bad because it threatens our future strengthens the mentality of instrumental utilitarianism: nature is valuable for its usefulness to us. Don't the planet and all its beings have a value in their own right? Or is the world, in the end, just a pile of instrumental stuff? If it is in one's self-interest to limit CO_2, it is even more in one country's, company's, or individual's self-interest to limit it less than its competitors. By appealing to self-interest and fear we strengthen the habits of self-interest and fear, which, let's face it, usually conspire to destroy the planet not save it. We will never increase the amount of care in the world by appealing to self-interest.

Invoking climate apocalypse devalues work that has little foreseeable relevance to climate change. Issues like poverty, homelessness, inequality, incarceration, racism, human trafficking, heavy metal pollution, GMOs, plastic pollution, and so on have a tenuous relationship to atmospheric health. Perhaps we should put all these causes on hold—after all, what will they matter if the planet becomes unlivable?—until we've solved the climate change problem.

This mentality is mistaken. The issues listed above have everything to do with climate, because the cause of climate instability is everything: every dimension of our separation from earth, nature, heart, truth, love, community, and compassion. If indeed self and world, humanity and nature, mirror each other and are part of each other, then it should stand to reason that climate instability will accompany instability in the social and political climate, and that imbalances in the natural realm will mirror imbalances in the human. Greenhouse gases are but a medium by which this principle operates.

A Story of Interbeing motivating earth healing neither contradicts nor depends on the story of saving the world from greenhouse gases. It supersedes it. The larger story of humanity in service to the healing and flourishing of Gaia is already bursting to take over. It is already the real motivating force behind a lot of work that people legitimize with climate narratives. Ecological restoration gets a boost in funding and attention if you make it about climate change, but as described above this is a dangerous strategy. Let's embrace the true motivation. Let's admit that we are acting from love of *this* forest, not for its instrumental role in global carbon sequestration. That we are acting from love of *this* soil, *this* lake, *this* estuary, *this* place, trusting that the health of one will contribute to the health of all, whether or not we can cite a climate argument to prove it.

A friend wrote to me about her involvement in the California Healthy Soils Initiative, "The initiative is about soil but it has to pretend to be about climate change to get funding, much like every environmental problem." It might get more funding that way, but such arguments often seem contrived. No wonder right-wing bloggers accuse climate change activists of having an ulterior "agenda." I think most of them *do* have an agenda—only it isn't to install a socialist One World Government or further the diabolical plots of George Soros.[1] It is to protect what is sacred to us. And so we use climate arguments for causes that aren't at heart about climate change.

Climate activists, listen. Understanding the role of ecosystems in maintaining climate equilibrium means that you can bypass skeptics' arguments

---

[1] If you have no idea what I'm talking about, read the comments sections of right-wing articles about climate change.

and build alliances by framing issues outside the carbon narrative. That's because you understand that any ecological healing will stabilize the climate too.

The same is true of social, cultural, relational, and personal healing. Everything is connected to everything else. While quantitative arguments can never demonstrate it, in our poetic hearts we know that the atmospheric climate somehow mirrors the political climate, social climate, and spiritual climate, and vice versa.

The Causes of Passivity

In most of the world outside the U.S., it is not the skeptics that pose the biggest obstacle to climate action, but rather the indifference of the general public and the political class. They profess to believe, but do they really? I am writing this at my brother's farm. What if someone came to me and said, "Hey, Charles, your four-year-old is outside wandering in that area where we saw a venomous snake!" And I said, "I believe you. I'd better do something about it and I will—as soon as I finish this game of Tetris." You would be justified in concluding that I didn't really believe the warning. Maybe I believed it was a milk snake we saw, not a copperhead. Maybe the person giving the warning has a history of false alarms. Whatever the reason, you could be sure I didn't really believe it, because if I truly believed my child were in danger I would drop everything else to protect him.

A majority of the public holds the opinion that climate change is seriously threatening civilization, but do they really believe it? Perhaps they are not so different from outright climate skeptics. The skeptics match their disbelief with the opinions they profess. The ostensible believer thinks he believes, but actually does not. Come on, do you *really* believe it? Or is it that you have episodes of despair in which climate change seems crushingly real, while other times you profess to believe but act as if the future of humanity were not at stake? Other environmentalists have raised this question, typically following with hand-wringing about how to cut through the denial and make people believe for real. Usually the strategy to do that has been to amp up the fear. I am arguing that a frontal assault on denial (whether the psychological denial of the average citizen, or the ideological position of climate skeptics) is unnecessary and has not worked. In fact, the more hyperbolic the headlines, the less effective they are.

For the last twenty years, 93 percent of all news in the "environment" category has been focused on climate change.[2] It seems nearly every article says either "See, climate change is really happening!" or "Such-and-such a hurricane or fire or famine was caused or exacerbated by climate change." Yet despite the crescendo of alarums, society as a whole still does not actually believe it. Quite the opposite: according to psychologist Per Espen Stoknes, "Long-term surveys show that people were more concerned with climate change in wealthy democracies 25 years ago than they are today. So the more science, the more IPCC assessments we have, the more the evidence accumulates, the less concerned the public is. To the rational mind this is a complete mystery."[3]

Stoknes explains the mystery in his book *What We Think about When We Try Not to Think about Global Warming.* Because the consequences of climate change are far off in time and space, people deprioritize it in favor of more immediate issues. For most people, compared to their mortgage payment or their teenager's addiction problem, climate change seems quite remote and theoretical—something that is only happening in the future or on the news. Even if one is intellectually persuaded of the reality and gravity of climate change, the felt reality is still "It isn't real" or "It's gonna be fine." Furthermore, Stoknes says, the climate issue is usually presented in doom-and-gloom terms that make people feel powerless to do anything about it, while at the same time feeling guilty about their inaction

---

[2] I totally invented that statistic. My point seems more persuasive with a number attached though, doesn't it? I'm sure if I picked the right database and methodology I could produce that number, or indeed virtually any number I chose. This exemplifies the obfuscatory power of numbers. We must always ask what hides behind them.

[3] Schiffman (2015).

and their complicity in the fossil fuel economy. This leads to various kinds of psychological denial to alleviate the guilt.

To Stoknes's point about the temporal and spatial remoteness of climate change, I would add another, more insidious form of distancing. The reliance of the climate narrative on global datasets and computer models creates a gap between cause and effect that can be bridged only by trusting the pronouncements of the scientific establishment. Even for those disposed to trust science, cause and effect are more distant than they are in the case of "logging is ruining the forest" or "toxic waste is polluting the river."

When we say that such-and-such a flood in Bangladesh or drought in Niger was worsened by climate change, people have to accept it as an article of faith, because Science Says So. Compare that to the water paradigm discussed earlier, in which the impending destruction of the wetlands of the Sahel by new dams will have devastating regional (and perhaps global) climate effects. A much shorter chain of causal argument is available here. Drain the wetlands, and the birds die, the soil bakes, the animals disappear, and the droughts intensify.

Around the world, deforestation, wetlands draining, industrial agriculture, hydroelectric dams, and urbanization make land vulnerable to catastrophic floods, droughts, and temperature extremes. All of these practices are addressable on the local level. The climate change narrative tends to make them seem inconsequential—a drop in the bucket of global emissions. It shifts attention away from the local devastation toward distant, perhaps hypothetical, effects.

The alternative frame I am proposing, focusing on local ecosystems, nullifies Stoknes's mechanisms of denial and paralysis. It addresses tangible damage in ways that bring tangible results. People cannot see changes in atmospheric concentration of invisible, odorless gases, nor can they be directly aware of distant effects on climate, but they can see (or feel the effects of) denuded hillsides, erosion gullies, smog, toxic waste, contaminated water, and so forth. They can also see the return of songbirds, the rising of water tables, the return of fish, and the clearing of air and water pollution where pro-environmental policies are implemented.

A problem remains, however. It is not only climate change whose effects are distant from everyday life—it is ecological destruction generally. This is especially true in the developed world. So far the elite nations are able to insulate themselves from the harm that ecological destruction causes. Therefore it seems unreal. The air conditioner still runs. The car still goes. The credit card still works. The garbage truck takes away the trash. School is open at 8 a.m. and there is food in the supermarket and medicine in the pharmacy. The routines that define normal life are still intact. If we wait for catastrophes to demolish them it will be too late.

As long as normal routines continue, most people will not be persuaded to take meaningful action. Persuasion does not penetrate deeply enough. No one is ever "persuaded" to make major changes in their life's commitments unless that persuasion is accompanied by an experience that impacts them on a physical and emotional level.

That is why, whether in the standard climate narrative or the more locally oriented ecological degradation narrative I'm proposing, we need to reach people in some other way. We need to pierce the perceptual, emotional, and systemic structures that separate people from their love for all beings of this earth.

Here is what I want everyone in the climate change movement to hear: People are not going to be frightened into caring. Scientific predictions about what will happen ten, twenty, or fifty years in the future are not going to make them care, not enough. What we need is the level of energy and commitment that we saw at Standing Rock. We need the breadth of activism we saw in Flint, Michigan, where everyone from yoga teachers to biker gangs joined in relentless protest against lead contamination. That requires making it personal. And that requires facing the reality of loss. Facing the reality of loss is called grief. There is no other way.

The Standing Rock action to stop the Dakota Access Pipeline wasn't framed around climate change at all (at least until white environmentalists became involved) but around protecting water and the integrity of indigenous sites, and not all water or all sites, but a specific body of water and specific sites, real places. Thousands of people, especially young people, braved long journeys and hostile conditions to participate. That is the kind of commitment we need to arouse in defense of the sacred, in defense of all beings of earth. It comes from beauty, loss, love, and grief.

Could we still drill new oil and gas wells, build new pipelines, open more quarries, and dig new coal mines if we came from a place of love for the earth and water around us? We could not, and anthropogenic global warming would be a moot question. True, the Standing Rock movement failed to stop the Dakota Access Pipeline, yet it revealed a tremendous latent power in that so many people were willing to go to such great lengths in defense of the sacred. What will be possible when that power is fully mobilized?

What would happen if we revalued the local, the immediate, the qualitative, the living, and the beautiful? We would still oppose most of what climate change activists oppose, but for different reasons: tar sands oil extraction because it kills the forests and mars the landscape; mountaintop removal because it obliterates sacred mountains; fracking because it insults and degrades the water; offshore oil drilling because oil spills poison wildlife; road building because it carves up the land, creates roadkill, contributes to suburbanization and habitat destruction, and accelerates the loss of community. Just look at photos of Albertan tar sand pits. Even if you know nothing about the greenhouse effect, the heart weeps at the toxic pits and ponds where pristine forests once stood. Or watch the *Gasland* films. Read about the oil spills that have devastated the Niger Delta. These immediate tragedies pierce straight to the heart, regardless of one's opinion about global warming.

From this vantage, we still seek to change nearly everything that the CO_2 narrative names as dangerous, but for different reasons and with different eyes. We no longer have to conjoin environmentalism with faith in Big Science and institutional authority, implying that if only people had more trust in the authorities (in this case scientific, but it extends to all the systems that embed and legitimize the institution of science) then things would be fine. You know what? Even if I were to accept the position of the climate skeptics, it wouldn't diminish my environmental zeal one bit. Awakening ecological consciousness doesn't require winning an intellectual debate with the skeptical forces. That isn't what will make people care.

By framing environmental issues in terms of CO_2, we distance people from grief and horror. Averting our eyes from the bulldozers toward graphs of CO_2 concentrations and average global temperatures, it seems perfectly reasonable to say, "Well, we'll offset that gas field by planting a forest. And besides, it's transitional until we get enough wind turbines operating."

Paradoxically, the CO_2 framing enables the continuation of the activities that are generating CO_2. On the global scale, any local power plant or city makes a negligible contribution to greenhouse gases. Any city could say, "We don't need to cut back on emissions as long as the rest of the world does." Any nation can say, "We cannot afford the economic cost. Let other nations make the cuts." The disputes that plague climate talks are inevitable when the problem and solution are framed in global, quantitative terms.

When we shift attention to palpable, local damage, such passing of responsibility to distant others is no longer possible. No one can say, "Let someone else preserve our beloved mountaintop. Let someone else preserve our beloved river. Let someone else preserve our beloved forest." We won't be mollified if the destruction of our favorite trout stream is "offset" by a reforestation project in Nepal. Not-in-my-backyard thinking, when universalized to an empowered citizenry, becomes not-in-anyone's-backyard.

Our family friend, the late Roy Brubaker, was a Mennonite minister in central Pennsylvania. He organized a highly successful watershed conservation campaign in his region, which is politically extremely conservative, by mobilizing the Rod and Gun Club. In the entire county it would be hard to find a Hillary Clinton voter, or anyone who would have lifted a finger had he framed the issue in terms of climate change. Yet, not only was the local watershed improved, with benefits downstream for the Chesapeake Bay, but if the living planet view I've advanced here is correct, the whole planet benefited as well.

Does deemphasizing the carbon narrative mean that business-as-usual gets a free pass? No. It is the contrary. As Wolfgang Sachs presciently observed, "Indeed, after 'ignorance' and 'poverty' in previous decades, 'survival of the planet' is likely to become that well-publicized emergency of the 1990s, in whose name a new frenzy of development will be unleashed."[4]

Protecting and healing local ecosystems around the world is much more disruptive to civilization as we know it than weaning ourselves off fossil fuels. Mainstream climate policy assumes that we can simply switch to renewable fuel to power industrial society and continued global economic development; hence the terms "green growth" and "sustainable development." The powers-that-be are quite comfortable with climate change when it is conceived in a way that gives more power to themselves, who are charged with, as Sachs puts it, "the Promethean task of keeping the global industrial machine running at ever increasing speed, and safeguarding at the same time the biosphere of the planet." This, he continues,

> ... will require a quantum leap in surveillance and regulation. How else should the myriad decisions, from the individual to the national and the global level, be brought into line? It is of secondary importance whether the streamlining of industrialism will be achieved, if at all, through market incentives, strict legislation, remedial programmes, sophisticated spying or outright prohibitions. What matters is that all these strategies call for more centralism, in particular for a stronger state. Since ecocrats rarely call in question the industrial model of living in order to reduce the burden on nature, they are left with the necessity of synchronizing the innumerable activities of society with all the skill, foresight and tools of advancing technology they can muster.[5]

Climate change portends a revolution in the relationship between nature and civilization, but this is not a revolution in the more efficient allocation of global resources in the program of endless growth. It is a revolution of love. It is to know the forests as sacred again, and the mangroves and the rivers, the mountains and the reefs, each and every one. It is to love them for their own beingness, and not merely to protect them because of their climate benefits.

---

[4]   Sachs (2010), 24. (This essay in the compilation was written before the 1990s.)

[5]   Ibid., 35.

The idea that deep and active care for the planet comes through experiences of beauty and grief, and not from fear of future ruin, might seem counterintuitive. Many people tell me they became environmentalists when they learned about the imminent, catastrophic consequences of climate change. Accordingly, we adopt the language of costs and consequences, hoping thereby to make others care about the environment.

But is that really why you became an environmentalist? The use of climate arguments to promote other conservation issues has a psychological counterpart in cultivating an image and a self-image of hardheaded realism, in which squishy nature-lover reasons give way to rational utilitarian ones. You can traffic in data about sea levels and economic losses and crop failure risks to disguise the truth: basically, you are a tree hugger. You are a whale lover, a butterfly gazer, a turtle caresser. Maybe you practice Druidic rituals or connect with the soul of Gaia in vision quests. The arguments you give about future impacts, 1.5 degrees or 2 degrees, meters of sea level rise, hectares of forest, energy return on energy investment for photovoltaics, methane clathrate release rates ... these legitimize your mushy tree-hugger sentiments. But this might be a Faustian bargain too, in which environmentalism accedes to the language of power, in exchange for its soul.

The bargain might be worth it if it actually brought the intended results. It hasn't. The ecological situation on Earth has deteriorated steadily, despite the adoption of data-driven models and the cost-benefit arguments that follow them. We have tried being reasonable. Perhaps it is time to be unreasonable. The lover does not need self-interested *reasons* to cherish his beloved. If we honor our inner nature-lover and speak from that place, others will hear us. Perhaps we have been speaking the wrong language, seeking a change of mind when really what we need is a change of heart.

Why Should I Love My Son?

Cam Webb, a rainforest ecologist, tells me, "The rich, lowland forests of Borneo are almost gone, with their giant trees (the tallest rainforest trees in the world are in Borneo), and many species of huge hornbill birds, and sweet-songed gibbons, and orangutans. When I first arrived in 1989, a green youth, I spent a full year living in the forest. During that year we would often climb to the top of the mountain, and look out into what felt like a thousand miles of unbroken rainforest. Now looking out, you can see that the park is an island in a sea of oil palm." When I met Cam, he was full of grief. "The rainforests I've spent my career studying are gone," he said. Is he sad because he performs an implicit greenhouse gas arithmetic in which Borneo's deforestation makes catastrophic climate change X percent more likely? Of course not. He loved those rainforests for themselves, not for their use.

At bottom, carbon arithmetic, even when extended beyond fossil fuel combustion to value the sequestration contributions of fish, grass, and trees, values things for the numbers they generate and not for themselves. Whether we value something for the profits it will bring or for the carbon it will offset, we are still instrumentalizing and objectifying it. The next step is inevitably to exploit and degrade it. Whether we do it to nature or to human beings, the results are ultimately monstrous, even if the initial intention is benign.

I am the father for four boys, including a four-year-old. Imagine I said to you, "Finally my older boys are grown, and now I've got another one. What a waste of time and money it is to feed him. I don't see a good reason why I should bother. Maybe I'll abandon him—what do you think?"

Imagine then that you replied, "Well, Charles, if you do that you might go to jail for child neglect. And even if you get away with it, he won't be willing to support you in your old age. Besides, what will the neighbors think?"

"You're right," I say. "I guess I'd better take care of him after all."

Obviously, if I have to ask why I should care for my own child, there is already a problem. No matter how strong the incentives and deterrents used to enforce care, I won't do as good a job as I would coming from love. I will do just enough to avoid negative consequences. I will do what it takes to satisfy the law and my neighbors. I might meet his every quantifiable need, if I'm closely monitored enough; I might even spend with him the required amount of "family time." But no list of quantifiable parenting standards can ever substitute for love. Will that family time be perfunctory or heartfelt? You cannot pay me or coerce me or scare me into actually loving my child. And if I don't actually love him, he will not thrive.

To make matters worse, by citing the law, the neighbors, and my old age as reasons to care for my son, you imply that if I could avoid bad consequences it would be fine to throw him out on the street.

Notice that the three reasons you gave me to care for my son mirror the ways we try to compel pro-environmental behavior. "You'll get arrested for child neglect"—we bring legal penalties to bear to deter pollution and other violations. "He won't take care of you in your old age"—we attempt to show that "green" policies are actually good for the corporate bottom line or the national economy. "What will everyone think?"—we appeal to PR considerations to make corporations and governments protect the environment.

The result of such inducements is often perfunctory compliance, evasion, exploitation of loopholes, and outright cheating. The corporation, government, or individual might abide by the explicit regulations while ignoring obvious harm that hasn't been prohibited by law or appeared on the radar screen of watchdog organizations. Rewards and threats don't produce real care.

I just looked up "biodiversity loss" on Google, and the number two result was, "How does biodiversity loss affect me and everyone else?" Essentially that webpage answered the question "Why should I care?" with "Because our health and livelihood are threatened."

Herein lies a problem: that answer also implies that if your health and livelihood are not threatened, then you needn't care. So, are they threatened? I think yes, but someone who subscribes to the story of technological utopia might say no. He might say that we will invent substitutes for everything the biosphere provides: synthetic food, an artificially maintained atmosphere inside bubble cities, and so forth. Moreover, even if you intellectually accept that biodiversity loss threatens human well-being, there is little in our lived experience to confirm it, since modern life so thoroughly insulates us from nature. How is your health right now? How is your livelihood? If you are sick and broke, can you honestly say that the extinction of the Bramble Cay mosaic-tailed rat or the Rabbs' treefrog had anything to do with it? Is saving the Yangtze finless porpoise going to help with the next rent payment?

When we propose the question "Why should I care?" and offer an answer, we have forfeited the argument. Caring about other beings, about life, about our planet is aboriginal to our humanness. To offer someone a selfish reason for caring is an insult. It says, "I know you. If it weren't for the threat to your wealth, health, or ego, you'd be just as happy to trample everything else for personal gain." Unfortunately, that is the Story of the Self that economics and genetics have offered, asserting that people are fundamentally motivated by self-interest. That way of thinking immerses us. Why are the birds singing? To mark territory and attract a mate. Why are the kittens playing? To practice hunting skills. Why are raspberries so delicious? To invite animals to eat them and poop out the seeds.

I find these answers dispiriting and, like the "Why should I care?" question, insulting to their subjects. I won't try to outline an alternative to neo-Darwinian genetic determinism in these pages, so I'll just offer a sponsoring principle: All beings yearn toward the exuberant expression of their life force. Birds sing much more than necessary; kittens play more than they have to; raspberries taste better than they need to. And you too, my friend, yearn to express your gifts in a beautiful way, more beautiful than necessary to secure a living.

The story we hold about another creates an invitation for the other to live into that story. Let us, then, invite each other into that inherent love for life that lies buried, however deeply, beneath the habits and beliefs of Separation. Let us create opportunities to act on that invitation. Because I know you—given half a chance, you will do it.

Nature Trafficking

Imagine for a moment that I had life-and-death power over you. Should I spare your life because you are more valuable to me alive than dead? You might be glad I did, but you will have no security, because what happens when the economics change? What happens when you are no longer valuable to me?

This is no idle speculation. Globally, life-giving resources go toward those who are "valuable" to the global economy. Those whose contributions do not translate into salable goods and services have trouble surviving. In conventional economic terms, they are indeed no more valuable alive than dead, and their ranks are growing. Only if we look at them through a nonfinancial lens, is their worth equal to anyone else's.[6]

The uniqueness and sacredness of any being disappears when it is reduced to a set of numbers.

The extreme case of human trafficking highlights the resulting degradation, which visits all of us in dilute form whenever we occupy the role of employee or consumer. When we are seen for our monetary value, our well-being ceases to matter except to the extent that it affects that value. The logic is apparent in arguments for employee health programs that cite the cost savings that will result from better health. Fine. But what happens when the costs outweigh the advantages of better health? The self-same logic says: sacrifice health. That's exactly what often happens when a company discovers a health hazard that would be expensive to fix. "We'll let that one slide."

Can we please understand that *this* is the revolution: to love all beings for themselves and not for their use? When we open to that, it isn't only our relationship to nature that will change. It also means the transformation of our economic system, which is founded precisely on the exploitation of human beings for profit; i.e., for their use. You probably don't like being treated that way, as an instrument of someone else's self-interest, as a consumer, as an employee, whose worth ends as soon as your money

---

6   Conventional economic theory equates economic value with value to society, through the following logic: Those whose contributions are wanted and needed will find that people are willing to pay for those contributions. Those whose contributions are unwanted will find no market. Those who make the most money do so because they create the most value. The flaws in that argument parallel the problems with any system of quantitative value. The metric used (money, CO_2, etc.) may not be an accurate or complete measure of value.

runs out or your productivity is exhausted. The beings of nature don't like it either. All expressions of the mindset of exploitation must change in tandem; each reflects and supports the rest. That is why all the revolutions under way today are the same revolution.

As with human trafficking, so with nature trafficking. Just as economic and political logic does with human populations, carbon accounting enables us to say, "This land is more important than that land. This species is more valuable than that species." The next step, of course, is to sacrifice the ones we think are less valuable, according to the numbers.

Quantification and monetization go hand in hand. Having evaluated something according to one metric, it is easy to map that onto another metric: money. Once we've equated green with low carbon, we can put a price on carbon to align green with money. This is the basic logic behind schemes to "monetize ecosystem services."

It is also the logic behind a genre of pro-environmental writing exemplified by this *Scientific American* subheadline: "Fish save the world billions of dollars in damages by helping store carbon dioxide in the oceans."[7] The article describes a study showing that fish in the high seas avert \$74–\$220 billion in climate damage per year. That's a lot more than the economic value of the fishing industry; therefore, the article concludes, we should change our fisheries policy.

Lucky thing for the fish that they are saving us money. Lucky thing for the employees that they are more profitable healthy than sick. Lucky thing for the honeybees that they provide such economically valuable services. But too bad for anything or anyone whose value registers low on our meter.

Do you know that feeling of enchantment on seeing a rare bird or on having a close encounter with an animal, seeing an eagle over the water, a whale spouting in the sea? Can you quantify how much poorer you would be without those beings? Come on, give me a number. Then we will know whether these are worth protecting.

---

7   Harball (2014).

In case you were wondering whether the oceans are worth protecting, the World Wildlife Fund has helpfully put a monetary value on them: $24 trillion.[8] No doubt it hopes to align economic incentive with ecological well-being: a laudable motivation. But think for a moment about the mentality this kind of valuation feeds. It suggests:

1. That money is a valid way to assess the value of something like an ocean

2. That we can and should make decisions about the planet based on the foreseeable financial gains and losses, and therefore ...

3. That if we could make more than $24 trillion (say, $48 trillion) by trashing the oceans, then we should do it

4. That it is possible to foresee and calculate the contribution of the oceans to human welfare in the first place—that our knowledge is sufficient to qualify us to even make this valuation

5. That we can separate out the oceans from the rest of the planet, as if they were a line item on a spreadsheet independent of the rest. So conceivably, we could compensate for the loss of the oceans with more of some other revenue stream.

6. That decisions about the oceans should be made based on the effect on human beings; that the oceans themselves and everything living in them have no intrinsic worth. What is important is their economic worth—their value to us.

Clearly, this mentality is part of the problem. At this very moment we *are* trashing the oceans for the sake of money. I do not know how many trillions of dollars we are making in the process, but when I read of ten thousand seals washing up dead on California's beaches, or hundreds of beached whales in New Zealand, or sea birds choked with plastic, or disappearing coral reefs, I know that however much we are making, it isn't enough.

---

8   Hoegh-Guldberg et al. (2015).

We have to understand that some things are beyond measure and beyond price. This conflicts with the reigning ideology of our time: science says that nothing is beyond measure; economics that nothing is beyond price. Accordingly, we (the dominant culture) have believed that by extending the scope and accuracy of our quantitative reasoning, we will through technology conquer the world, and that by extending the domain of market relationships, we will maximize the efficient production of wealth.

Why then, even as our technologies of control grow more powerful and precise, does the world seem to be spiraling *out* of control? Why then, even as global GDP reaches new heights, do we experience greater and greater poverty—a poverty from which even the financially rich are not exempt? It is because of what is left out of our measurements, the hard-to-measure and the unmeasurable: beauty, joy, suffering, purpose, pain, sacredness, fulfillment, play ... and the sight of seals on the beach, even if they were useless for any other purpose. Yet these are what make life rich.

Ironically, the mindset of instrumental utility, which evaluates all things for how they benefit ourselves, doesn't even benefit ourselves. That is rather hard to explain in the Story of Separation, except perhaps by saying we have to do a better job at it. We need to be more clever and less shortsighted in our exploitation of everything else for our benefit. But in the Story of Interbeing it is obvious why. In a world of intimate relatedness, harm to one is harm to all. Our efforts at control will always bear a limit; our efforts to measure and predict will never be complete.

Numbers have their place, but if we are to preserve the things on this planet that are beyond price, we cannot rely on math to do it. We cannot scare ourselves into compassion, imagining that properly quantifying the blowback will deter us from wreaking further harm. (Fear for one's self-interest is what thwarts compassion in the first place.) Nor can we bribe ourselves into love, hoping that we will finally take care of our oceans if only we realize how much money we'd save. The pecuniary mind won't save us from the destruction generated by the pecuniary mind.

When we appeal to utility as a way to promote sustainability, we implicitly affirm the normalcy and rightness of making decisions on the basis of utility. That is counterproductive, because most of the time, whether you are a business or a consumer, what is of immediate, calculable utility to you does harm to the planet. Self-interest as our culture constructs it says to the mining company, "Strip-mine that forest." It says to you, "Buy that smartphone made from the strip-mined minerals." It says, perhaps, we can plant another forest somewhere else to compensate for the lost carbon sink. It also entices us to just go ahead and exploit, because what really contributes to my own personal utility is for someone else—not me—to forgo the minerals, the profits, the smartphone.

If a forest is sacred to you, then how much would I have to pay you to cut it down? No amount would be enough, just as no amount of money would be enough to induce you to offer your mother or child for extermination. When we translate the value of a forest or other ecosystem into a carbon sequestration number, its value becomes just that, a number, something finite. The forest becomes expendable, so long as something of greater value can be substituted.

At best, financial arguments allow the bean counters (inner and outer) to relax their guard and give us permission to act from our love of Earth. "It's okay. It makes economic sense too." Unfortunately, they also perpetuate the notion that the ecosystem is, fundamentally, a source of "services"; that the planet is here for us, valuable for its use to us, and not in its own right. The ecological revolution must go deeper than that.

Rights of Nature

The revolution is love. It isn't about more cleverly valuing and utilizing nature. It is about genuine respect for nature, which can come only by seeing it as a being in all fullness, and holding it sacred. Where is the sacredness when we have reduced it to a finite value? We need a better reason to care for the world, a truer reason. We need to connect with a source of motivation that isn't even reasonable.

In writing this book, I was tempted (and advised) to avoid saying things like "The earth is alive and sentient." Such statements exclude me from consideration by policymakers, who need arguments framed in rational terms. But can we ever reason our way to love? "Rational" in this context is usually a code for utilitarianism. Since when is love rational? The truth is, we love the earth for what it is, not merely for what it provides.

I suspect that even the most hardheaded environmentalist, who derides the Earth-is-alive crowd most vociferously, harbors a secret longing for the very object of his contempt. Deep down, he too believes the planet and everything on it is alive and sacred. He is afraid to touch that knowledge even as he longs for it.

This person is also me. The idea of a living, sentient Earth attracts me and repels me both, mirroring the polarity of opinion I observe at conferences between the nuts-and-bolts and spiritual factions. Accusations of "naive!" "softheaded!" and "unscientific" rattle around in my own brain, expressing a hurting thing within. Maybe if I join the ranks of the critics and turn the criticism outward, accuse others of ignoring science and indulging in fuzzy thinking, I can find some temporary relief. But it would be more honest of me to embrace my irrationality. And it might be more inspiring to others too, to invoke in them the same biophilia I know within myself.

The idea that our planet is alive, and further, that every mountain, river, lake, and forest is a living being, even a sentient, purposive, sacred being, is not a soppy emotional distraction from the environmental problems at hand; to the contrary, it disposes us to feel more, to care more, and to do more. No longer can we hide from our grief and love behind the ideology that the world is just a pile of stuff to be used instrumentally for our own ends.

Given how central instrumental utilitarianism is to the world-destroying machine, the environmental movement must take care not to reinforce that story with its rhetoric. It must inhabit, enact, and propagate a different story: of care, of beauty, and of love. This does not mean that it should ignore the consequences of ecocide for human beings—after all, we are also among Gaia's beloveds—but that it should avoid elevating those arguments to primary status. Yet this is the language, nearly the exclusive language, of "serious" policy conversations about climate and other environmental issues. It hasn't been working. Maybe we should try once again the language of love.

By denying the nonhuman material world the qualities of a lovable self, we make nature and the material world unlovable. If at bottom the world is composed of a bunch of generic, purposeless particles governed by impersonal, random forces, what is there to love? Locutions such as "natural resources" and even "the environment" foster this kind of ideological separation. Compassionate love comes from the realization that you are a self, just like I am. A child looks up at the sun and knows it looks back at her. Then we grow up and know better; we dismiss those perceptions as a childish anthropomorphic projection; the scientist does the same when he or she claims that only humans possess the full degree of consciousness, agency, intention, desire, and experience of being; that animals possess these things, perhaps, to a lesser degree—the "lower" the animal (i.e., the more unlike ourselves) the less; that plants possess only a rudimentary amount of them if at all; and that these qualities of a self are surely absent from rivers, mountains, soil, water, and rocks. But intuitively, we, like the child, like older cultures, know better. We know that the whole world that environs us is a self in all fullness, and so is every part of it.

While money is insufficient to reflect the value of that which is beyond price, we do have another instrument of human agreement to apply here: law. The growing Rights of Nature movement seeks to secure legal status for nonhuman beings; so far, Bolivia, Ecuador, and New Zealand have written these rights into law. The earth rights lawyer Polly Higgins has campaigned to extend them globally by adding ecocide to the list of crimes against peace alongside genocide, war crimes, crimes of aggression, and crimes against humanity, putting it under the jurisdiction of the International Criminal Court. This would elevate interbeing to something more than a personal philosophy or religious orientation. It would enshrine it as a foundational principle of a different kind of society.[9]

Once upon a time, science saw notions like the personhood of nature ridiculous. Although the science is changing (for example, a growing contingent of biologists is seriously considering the possibility of plant intelligence), to this day many would feel vulnerable to accusations of fuzzy-headedness if we said, "Who cares about the costs and benefits. Let's save this forest just because we love it. Let's save it because it is just so beautiful."

That is not to say we should never cut down trees. It is to say that such an act should not be facilitated by an ideology that holds trees—and all life—as anything other than sacred. When we see the forests in terms of board feet or timber value, when we see the oceans in terms of tons of protein or dollars' worth of fish catch, when we speak of nations as "economies" and people as "consumers"; when we see a place as a source of iron ore or bauxite or gold, when we see these minerals as nothing but minerals, randomly deposited and unrelated to the processes of life around them, when we see a forest or peat bog in terms of its carbon sequestration potential, then we are seeing Earth as a machine, not an organism, dead and not alive.

The reason our current system of material production kills the world is that it starts by seeing the world as dead. What then is there to love?

---

[9] "Rights of Nature" may not be the best term to confer legal personhood onto beings of nature, since the notion of rights takes as elementary the individual and the state. For indigenous and other community-based cultures, "rights" is not a coherent concept. We might be cautious about extending the notion to the nonhuman world. An alternative might be "responsibility to nature." The important thing is to somehow encode nonhuman personhood into the framework of agreements and narratives that we call law.

7

The Revolution is Love

In a Rhino, Everything

A couple years ago I received the following email from a young woman, a student at an elite law school.

I don't cry very often. But this week I cried twice. For the rhinos. It breaks my heart that they're going extinct. In order to make myself feel better, I try to intellectualize this. It's totally irrational, I say to myself, to be sad for the rhinos. Why not be sad for the fairy shrimp, going extinct right here in Southern California?

There are so many things to be sad about: police shootings, for example. Right now I'm writing a memo on what constitutes excessive force during arrest and when you type in excessive force and qualified immunity to Westlaw, more than 600,000 cases come up. These cases represent a tiny fraction of incidents with police brutality; far more go unreported or are never litigated. We have a police violence epidemic in this country. I could be sad about that. And here I am reading these cases—and it's awful (the tasing, the shooting, the beating, the pepper spray, the long term injuries, the easiness of getting out of excessive force charges) and I never cry.

And then I read some articles about the last, aging, white rhinos in zoos around the world and I fall to pieces. How can we have failed so badly? And you're right Charles, it's grief for the dying biosphere (I have long since stopped equating the environmental crisis with global warming, and I hate it when people do that).

There's this kid in my class who really gets under my skin. He says annoying things like, "I love it when I see pictures of McDonald's in other countries, or African kids wearing Nikes, because it's like we've won. Our culture is supreme." I gave him a look when he said this. And he knows how I think because we've had conversations, so he said, "I can't help it, I'm pro-American." And I said, "I'm pro-biosphere." And he says, "I think we should only keep the animals that we need to survive." And I'm so shocked by this stupidity that I'm rendered speechless. I literally couldn't talk to him for a few minutes. I didn't want to talk to him. I felt a little nauseous. Finally, I said, "I don't think that's possible." And he said, "Well, we can try." Like it's a good thing to try for. He gives me a panicky feeling because I think what if he's right? What if the future just contains concrete with cows, pigs, chickens, and their shit? What would we do with all their shit? (Previously he has told me that he could never care for an animal, that an animal's suffering has no effect on him.)

I'm really trying not to other him. I sat next to him in class this semester because I know I have something to learn from him. I try to be kind to him, even though the things he says make me ill. And it's not from a place of moral purity either. I'm trying to understand this behavior, this kind of thinking, because if I never understand it I'll never be able to confront it in a meaningful way. It's a challenge, though. Sometimes I feel my innate snarkiness rising to the surface, but I know this is just a defense mechanism on my part. Any suggestions?

The scariest thing about this kid is that he's totally pro-carbon controls. He believes in global warming and that it's a threat and that we should do something about it. I would prefer a climate change denier with a love for animals. Really, I would.

There's something, though, besides grief. The grief is compounded by that horrible sense of helplessness. I feel like I have absolutely no control over the fate of the rhino. I do my work, you know? I made all A's last semester ... I'm disciplined. I'm studious. But I'm not doing anything real.

Like this young woman, I do not know why some tragedies penetrate me with grief while others do not. There are endless things to weep for. Because we cannot weep for each one that comes across our awareness, we might form emotional calluses just in order to function. And then from time to time something pierces those calluses, and all the other unmourned tragedies follow it through the breach. Sometimes, therefore, it is a seemingly tiny thing that brings me to tears or heart-wrenching agony: a parent shaming a two-year-old child. Or it could be a heartbreaking injustice visited on the innocent: a child marooned in this country when her parents are deported. Or it could be a single incident of brutality out of millions that gets under my skin. Each of them represents the rest. In fact, each contains the rest. Next time you travel to another planet and see caged wild animals there going extinct, you will know that planet also warehouses its elderly in nursing homes. A world in which the last white rhinos are aging in zoos is also, necessarily, a world of incarceration, war, racism, poverty, and ecocide. It is impossible for one to exist without the others. All are part of the same unholy matrix.

Because each of these contains the others, when we grieve one of them we grieve them all. It doesn't matter if it is the rhinos or police brutality that pierces you. They are all expressions of the same underlying mythology: the story of a discrete and separate self in a desacralized world that is other.

If you take for granted a universe of standard building blocks, devoid of the qualities of a self, devoid of an inherent intelligence or evolutionary will, then our license to manipulate nature and materiality suffers no limit except for that posed by perverse unintended consequences that we can, in principle, predict and control with just a little more information and technological know-how. Why not, then, keep only the animals that are useful to us? In the Story of Separation, we are fundamentally separate from the rhinos. What happens to them needn't affect us, outside the realm of mushy sentimentality.

The same goes for the biosphere as for the rhinos. In the Story of Separation, what happens to the biosphere needn't affect us, except as a temporary practical matter pending the technology to make us independent of nature. That is the world of concrete and pig shit that my friend dreads.

Here is why her observation that she would prefer an animal-loving climate change denier to this person rings true. Love violates the Story of Separation. Love is the expansion of self to include another, whose well-being becomes part of one's own. The healing of our planet will not come without love for our planet. The animal-lover, at least, is on the right track.

If we want to change the minds of people like the woman's classmate, head-on debate isn't going to work. No one can logically persuade somebody to fall in love. We might be able to convince them to support one policy over another on utilitarian grounds, but engaging the planet as an instrument of our utility is what has gotten us into this mess to begin with. It reminds me of the "pragmatic" opponents of the Vietnam War and Iraq War who didn't question war as a tool to promote American interests (nor did they question the concept of American interests), but who merely said that this particular war wasn't working. The door to more war remained open. Similarly, when we say, "Let's stop using fossil fuels or we're screwed," and adopt anthropocentric interest as our primary argument, there is little to say for the rhinos. Will we be "screwed" if they go extinct? Probably not. And so we proceed toward a world of concrete and shit, with maybe a few parks for aesthetic relief.

Why would this man and millions like him be attracted to the Story of Separation that seeks to exploit and manipulate the world? Maybe it has something to do with he himself feeling like an instrument, exploited and manipulated. He is in the same position that he wishes to put the animals and the planet. Lacking real sovereignty, he longs for a sense of control. Humanity (as a proxy for the self) being in control of things feels good to him. Not to psychoanalyze the poor guy, but if we are serious about changing the beliefs that drive ecocide (rather than gaining the psychological gratification of winning an argument) it is important to understand the experience of life behind those beliefs. Ideology and psychology are inseparable.

I think this young woman is therefore on the right track, showing him kindness while not allowing herself to be dominated by him. In a worldview of winning and losing, no one will go out of their way to serve your interests unless you dominate them, force them, pay them. In its extreme, that world has no love, no real kindness, no generosity that isn't a device to get more. That is why unforced kindness and generosity have the power to puncture the Story of Separation.

The kindness my friend shows her classmate and the desire to understand his experience of the world translate onto the level of systems and politics. What is the story our opponents stand in, the perpetrators, the ones we want to blame? What kind of life experience attracts them to that story? What are the secret ways that it lives in ourselves? When we know what it is like to be them, we will be far more capable of disrupting the narrative armature of the world-destroying machine. This is called compassion. It isn't a substitute for strategy and action. It illuminates new strategies and makes action more effective, because we can target them at the deep causes rather than forever battling the symptoms.

What is it like to be a rhino? To be a policeman? A corporate executive, a terrorist, a killer? What is it like to be a river? These questions arise naturally in the Story of Interbeing, which holds us as interdependent on every level, even that of basic existence. They are not mere psychological questions; they are also economic and political questions, because it is these systems that generate much of our experience of life.

The lens of interbeing also relieves the helplessness the woman speaks of at the end of her email. Even as the crises of the world each contain the others in an "unholy matrix," the same is true for the responses. To respond to any is to respond to all. I imagine myself talking to a rhino in a cage. She asks me, "What were you doing with your life, while I was going extinct?" If I answer her, "I was working to save the coral reefs," or "I was helping to stop the navy from using whale-deafening sonar," or "I spent my life trying to free men from death row," then she is satisfied, and so am I. We both know that, somehow, all of these endeavors are in service to the rhinos too. I can meet her gaze without shame.

This is something that the entire spectrum of climate change positions, from skepticism to catastrophism, misses. A world in which babies are separated from mothers at birth, in which children are medicated to make them pay attention in school, in which we drain swamps and discharge toxic waste, in which human trafficking runs rampant, in which animals are confined in feedlots, in which punishment is mistaken for justice, in which wealth concentrates in ever fewer hands, and in which people hate each other because of the color of their skin, is necessarily a world where the climate is spinning out of balance. And these are not just signs, they are causes. That means that someone working to end the criminal punishment system is also helping to heal the climate. The causal link between criminal punishment and climate is probably beyond our understanding, but somehow each recognizes the other as an ally. Only in the disconnected worldview that spawns carbon reductionism can we think that climate change is somehow separable from everything else I have named.

In the Story of Interbeing, what happens to anything happens in some way to everything. We are free then to listen to what calls forth our passion, our care, and our gifts, whether the need that calls them seems large or small, consequential or invisible. Because each contains all, we can be peaceful in our fervor and patient in our urgency.

The Concrete World

We can be peaceful in our fervor and patient in our urgency. We let in the grief, and compassion and clarity follow it in. We stand in awe of the intelligence that weaves it all together and orchestrates the mysterious causal pathways that link the rhinos to the prisons to the corals to the cancer wards. But that does not answer my friend's trembling worry: What if he is right? What if the future just contains concrete with cows, pigs, chickens, and their shit?

In this book I have referenced the idea that human well-being and planetary health are inextricably connected, and that no such future is possible. I will now explore the opposite idea: that human ingenuity is unlimited, as is our capacity to replace ecosystem services with technological substitutes.

In other words, what if the version of the Story of Interbeing I've given you is wrong? What if we can, in fact, forever insulate ourselves from the effects of our actions? What if the techno-optimist position on the spectrum of climate change opinion is right—that climate change is just a technical hurdle in humanity's race to its glorious destiny?

While it is fashionable to attribute the collapse of past civilizations to ecological degradation, these narratives are subject to critique. Consider for example the famous case of Easter Island, whose transition from a wildlife-rich, forested paradise to a treeless desert island devoid of any large fauna is used as a cautionary object-lesson in population overshoot and ecological destruction. Jared Diamond, who popularized this narrative in his book *Collapse,* chronicled the descent of a highly developed civilization of more than fifteen thousand people into poverty and cannibalism, following their hunting to extinction the island's land and sea birds and cutting down the trees to fuel their obsession with building their famous stone monoliths.[1] The parallels to our own civilization are obvious.

This narrative has come into question recently by Terry Hunt and others, who argue that it was rats (which traveled to the island with the Polynesian colonizers) that were largely responsible for the deforestation, and that the human population never reached fifteen thousand but remained fairly constant until the arrival of Europeans.[2] In other words, ecological collapse was not followed by social collapse. To the contrary: despite ecological collapse, the society maintained its numbers and social cohesion on its desolated landscape. They spread broken volcanic rocks whose gradual breakdown fertilized vegetable gardens, they ate rats for protein, and they enjoyed enough abundance to continue building stone monoliths. Naturalist J. B. MacKinnon notes that population dropped and the culture unraveled only after first European contact brought lethal new diseases to the island.[3] At first contact, the locals were more interested in trading for hats than for food or other "necessities."[4] This was not a desperate society.

---

[1] Diamond (2005).

[2] Hunt (2006).

[3] MacKinnon (2013), 199.

[4] Ibid., 198.

History is often a projection screen for contemporary prejudices. In an era when we fear ecological collapse, it is natural that we would view history through that lens. Historical examples can also be used purposely to intensify the alarm that environmentalists hope will motivate change. I think, though, that the change we need must come from a different place than self-interested alarm.

For me, the implications of the revisionist Easter Island narrative are more chilling than the possibility of ecocide-driven collapse. It suggests that a concrete world of "cows, pigs, chickens, and their shit" might be possible after all. As MacKinnon puts it in his poignant and insightful book, *The Once and Future World,*

> *What the Easter Island stories represent are the two possible end points for our global culture if we continue on our current course toward an ever more simplified and degraded natural world. In the first telling, the fates of nature and humanity are entwined and both go down together in a social and ecological catastrophe. In the second telling, human and non-human life take different paths. The planet's ecosystem is reduced to a ruin, yet its people endure, worshipping their gods and coveting status objects while surviving on some futuristic equivalent of the Easter Islanders' rat meat and rock gardens.*[5]

This passage evokes a nightmare world where the entire biosphere has been converted to a giant feedlot and industrial park, where we manage the planet like a machine with technological tweaks to its gross material components, where no species exists that has not been turned to human purposes. It is a world wholly toxic to life except within artificially maintained enclaves. It is a world of vat-grown meat, computerized hydroponic greenhouses instead of farms, algae pools for oxygen, carbon-sucking machines to regulate the atmosphere, desalinization plants, climate-controlled air-filtered bubble cities, and a planetary surface converted to one huge mine and garbage dump. In that world, human life becomes entirely dependent on technology, as we retreat from the ugliness we have wrought into an

---

[5] Ibid.

artificial or even a virtual environment. Can you say this isn't already under way? This is not a world I would want to live in. No one would, yet for thousands of years, humanity in aggregate has proceeded choice by choice, step by step toward a Concrete World. I would like to dismiss it as impossible on ecological grounds, but what if it *is* possible? What if, instead of being compelled to reject it, we must consciously choose a different path?

Climate activists are fond of saying, "We are going to *have* to change now." Maybe the significance of climate change is not "Change or perish," but an invitation to reorient civilization toward beauty rather than quantity. Starkly confronting us with the results of our power, it asks, "What kind of world do you want to live in?"

Whether or not endless technological adaptation to an ever-more-degraded ecosystem is actually possible, the *perception* that it is possible indeed presents us with the necessity of making a conscious choice. If ecological degradation had the power to force us to choose a healing path, it would have happened already. Therefore, that choice to take the healing path will have to be made on some basis other than compulsion. It will not come through fear of personal or civilizational extinction.

I want to repeat: if ecological degradation had the power to force us to choose a healing path, it would have happened already.

Is the choice to heal ever really forced on us? Some people quit smoking on the first diagnosis of lung disease; others persist in smoking through their tracheostomy hole even as lung cancer wracks their body. What is happening when we reach that critical moment where we hit bottom, where the old life becomes intolerable? When do we say, "Enough. I'm out of here"? When do we finally quit that job, leave that relationship, take that journey, quit that addiction, release that grudge? Usually a course reversal toward wholeness is sparked by some kind of crisis, but it is not guaranteed by one. Each crisis, each tragedy, each new injury or loss is an invitation onto a different path. It is up to us to accept that invitation.

As the deterioration of the biosphere proceeds, we will surely face many crises, tragedies, and losses. If fear of further loss is not enough to change our course, what is? The dominant environmental narrative, especially when it comes to climate, is based on fear of consequences for humanity. What do we choose from when we don't choose from that fear?

Most people will offer love as the antipode of fear. I'm wary of this formula, which veers close to replicating the familiar paradigm of good versus evil. Fear is not always a bad thing; sometimes it can heighten one's wakefulness and focus, and catalyze action. That action might be in service of those we love; it needn't be in service of self-preservation. We care about what we love, even when that caring in no rational way contributes to our measurable benefit. Sometimes we even sacrifice our lives for what we love. Love makes us care more passionately than self-interest can, and, as Dr. Seuss put it in *The Lorax,* "Unless we begin caring a whole awful lot, nothing is going to get better. It's not." Nature might not save us from ourselves.

So the care we need to live in a more beautiful world comes from love. But how does love awaken? One way is through loss, grief, and the realization of death. When a friend or family member falls ill or has a near brush with death, or enters into the dying process, the reality of their preciousness overcomes my holding patterns and opens me to deeper care. Unfortunately, these holding patterns have a powerful ally in modern society's denial of death—its fetish for youth, preservation, and growth. Denial of death holds life at bay too. It usurps love and enthrones the pretender we call ego. Modern civilization upholds an equivalent pretense in its ideology of human exceptionalism: that human beings and human society can be exempt from limits. The untrammeled growth of the separate self, whether the personal self or the collective human self, is inimical to love. Therefore, death, loss, and grief are love's allies.

Another partner of love, in its awakening and its expression, is beauty. We fall in love with what is beautiful, and we see beauty in what we love. Let us consider beauty then to replace utilitarian benefit as the motivation and aim of humanity's relation to the world. Whether or not a world of rats and concrete is survivable, it is definitely not as beautiful as the "once and future world" that MacKinnon describes.[6] Explorers and naturalists of previous centuries give staggering testimony to the incredible natural wealth of North America and other places before colonization. Here are some images from another book, Steve Nicholls's *Paradise Found*:

Atlantic salmon runs so abundant no one is able to sleep for their noise. Islands "as full of birds as a meadow is full of grass." Whales so numerous they were a hazard to shipping, their spouts filling the entire sea with foam. Oysters more than a foot wide. An island covered by so many egrets that the bushes appeared pure white. Swans so plentiful the shores appear to be dressed in white drapery. Colonies of Eskimo curlews so thick it looked like the land was smoking. White pines two hundred feet high. Spruce trees twenty feet in circumference. Black oaks thirty feet in girth. Hollowed-out sycamores able to shelter thirty men in a storm. Cod weighing two hundred pounds (today they weigh perhaps ten). Cod fisheries where "the number of the cod seems equal that of the grains of sand." A man who reported "more than six hundred fish could be taken with a single cast of the net, and one fish was so big that twelve colonists could dine on it and still have some left."

I used the word "incredible" advisedly when I introduced these images. Incredible means something like "impossible to believe"; indeed, incredulity is a common response when we are confronted with evidence that things were once vastly different than they are now. MacKinnon illustrates this phenomenon, known in psychology as "change blindness," with an anecdote about fish photographs from the Florida Keys. Old photographs from the 1940s show delighted fishermen displaying their prize

---

[6] The reader might object that aesthetic sensibilities vary; that some people find glass skyscrapers more beautiful than forests and waterfalls. I myself find certain skyscrapers beautiful (though rarely any built after 1950). One might ask: is it really beauty that draws people to environments of steel, glass, and chrome, or is it security? And do they know what they are missing?

catches—marlins as long as a man is tall. When present-day fishermen see those pictures, they flat-out refuse to believe they are authentic.

Human beings tend to be blind to gradual changes in their environment, assuming that the way things are right now is how they always have been and always will be. We do not miss the former beauty of the world, he would say, because we have never known it.

I am not so sure that we don't miss it. I think we do miss it, but we don't know what it is we are missing. We feel a void, a sense of poverty, a hunger for something unidentifiable. Transferred onto money or consumer items, that hunger drives continued cycles of destruction. Transferred onto drugs, gambling, alcohol, it drives the unsolvable social problem of addiction. Perhaps the degradation of nature is not without its consequences after all.

Even a taste of this lost abundance nourishes me deeply and points to a kind of wealth that may one day be restored. Once I went swimming on the north coast of Scotland, and a seal swam over to check me out, popping its head out of the water in an aspect of comical curiosity. That image still enriches my brain. On my brother's farm in June I spend long minutes gazing at the lightning bugs, sparkling like Christmas decorations in numbers I haven't seen since my childhood. They make me feel at home in the world.

When I lived in Harrisburg I visited a greenbelt near my house every day. Although the land was enclosed by the city, and the stream tainted with leaking sewage, and the woods infested with ticks and poison ivy and drug deals at night, still I found special spots where I would go watch the play of the minnows, and stand where the birds found refuge. Without that nourishment I doubt I'd be writing this today.

Despite all that has been lost in our progress toward a concrete world, much beauty remains. The earth is still alive. Now is the time to choose life. It is not too late.

Most people would acknowledge a feeling of loss at the thought of a world without elephants, rhinos, or whales. But, the cynic might say, we'll get used to it and not know that anything is missing, just as you probably are not mourning the loss of the Pyrenean ibex or the hundreds of nameless species going extinct every year. However, in the Story of Interbeing, where self is relationship, each extinction impoverishes the web of relationships on Earth that includes ourselves; it shrinks us and simplifies us. Extinctions are the end result of an ideology that makes other beings into less than full beings and excludes them from the circle of self. First, they are cast out of full existence via our belief system; in the end, the casting out takes irrevocable physical form. First, the mythology of separation isolates us from our companions, who are really part of ourselves; then those companions perish forever.

This impoverishment goes beyond outright extinction. Many species, while not entirely extinct, persist as remnant populations on tiny fragments of their former range. Thus they recede from our lived experience. Moreover, modern people live almost entirely in a realm of products, media, and the indoors, estranging them from the life forms remaining in their ambit. I cannot identify the name and likeness of more than ten bird species from their songs. Can you? I hope you can, but I think most in my culture cannot. This degree of alienation is normal now.

One consequence of this is an ever-growing loneliness, an ache that nothing in the indoor world, manufactured world, or digital world can assuage. We miss the complement of our relationships in all their diversity. Standardized, digitized, or abstract relationships do not nourish full beingness. Surrounded by standardized commodities, visiting public spaces filled with strangers, interacting increasingly through the internet, and distanced from intimate relationship with nature in a world of climate-controlled houses, packaged food, and machine-mediated labor, we are poor in our very existence. Do we still survive? Yes. From the perspective of the Story of Separation, we continue to exist. But it is a partial, anemic existence. For the interconnected self, existence is not a binary yes/no. Existence admits to degrees that depend on wealth of relationship.

I think that "Could we survive in a ruined world of synthetic food and concrete algae pools?" is the wrong question. Better questions might be

"What will we become?" "Who do we want to be?" and "What kind of world shall we choose?"

The climate crisis and general ecological crisis may not be about the survival of our species at all. It may be an initiation into a new orientation altogether. The question then becomes not whether we can survive, but how we want to live. It becomes no longer how to achieve sustainability, but what we want to sustain.

The Conditions of Our Choice

Let us grant for a moment the premise of the technological fixers, that says climate change is not a problem because human ingenuity is limitless. If so, if we can manifest anything we put our minds to, then why settle for a world that grows uglier and more degraded with each passing year? And why settle for the inner desolation that accompanies it? Yes, perhaps we could use technology to compensate for the loss of ecosystem services. And yes, perhaps we could remediate the corresponding inner loss with technology too, with psychiatric medication, with "content-rich" virtual realities to compensate for the impoverishment of outer reality, with a profusion of diversions and stimulation to assuage the aesthetic, sensual, and psychic hunger caused by the depletion of the natural world. Perhaps we could.

But even if we could, we don't have to. We could instead devote this "limitless human ingenuity" to the wholeness and beauty of the entire world, applying what might be called "technologies of reunion" to the restoration of inner and outer landscapes. Granted the prodigious power of the human will, we might change the question I just asked, "Why settle for a world that grows uglier and more degraded with each passing year?" into another: "Why *have* we settled for a world that grows uglier and more degraded with each passing year?"

If we cannot answer that question, and if we cannot change the conditions of that choice, then there is no hope that we will reverse course. No hope at all. We will continue to settle for what we have always settled for.

The conditions of our choice to settle for a degraded world are so ubiquitous and unremitting that we take them for reality itself. Together, they weave the mythology and experience of Separation in which we live.

I have already described the basic metaphysics of the Story of Separation: the discrete and separate self in an objective universe that is Other, populated by impersonal forces, generic bits of matter, and competing other selves. Here are some of the means by which it results in ecocide:

Through the present system of money and property, which, as I lay out in *Sacred Economics*, reifies and justifies the ideology of competing separate selves. Usually, discussion of the influence of money on the environment goes only so far as to blame corporate greed, government corruption, and irresponsible shopping habits. Outside of the Left intelligentsia (and all too rarely even there), we seldom see a trenchant explanation of how the deep structure of capitalism as we know it makes continued ecocide inevitable. This is such an important matter that I will devote a full chapter to it, because absent that understanding, we will fruitlessly pursue strategies of sustainable development, blind to its inherent contradiction.

Through ideologies of reductionism and linearity that cause us always to underestimate the consequences of our actions. The conception of nature as a fantastically complicated machine obscures its wholeness and the interrelatedness of its parts. We know in a human that damage to one organ or tissue reverberates throughout the entire body, but only recently has the dominant civilization begun to appreciate that this is true for the ecological body as well. Damage in one place, the extinction of one species, the draining of one mangrove swamp, is impossible to contain, but generates distant effects that require yet more intervention. For example, the mechanistic mindset says if insect damage to crops is the problem, then insecticides are the solution. And if the insecticide kills nontarget species that kept a fungus under control, then the solution is a fungicide. And if the fungicide damages mycelial networks that maintain soil integrity and water retention, the solution is irrigation. And when irrigation and chemicals exhaust or poison the aquifer, the solution is to pipe water in from somewhere else. On and on it goes—a series of technical fixes that postpone the consequences of the damage indefinitely into the future, and that distance effects from causes.

In other words, we have chosen to continually degrade the biosphere because we don't know what we are choosing. Not knowing that Earth is an interconnected living body, we think we can isolate and contain the damage. We are mystified when it erupts somewhere else in changed form, its cause perhaps unrecognizable. Trapped in linear thinking, we again search for the most proximate, direct cause. Colony collapse disorder is killing the honeybees. Why? We try to find the cause, the pathogen, the enemy—something to fight against. Linear thinking is war thinking. Useful and appropriate sometimes, when exercised in ignorance of nonlinear feedback it results in an endless war against enemies created by the previous war. (Obviously this pattern holds in politics as well as in our relationship to the natural world.)

Today we are in an ideological transition zone between linear control-oriented thinking and nonlinear systems theoretic thinking. In genetics, for example, the old dogma of "one gene, one trait" has completely disintegrated as it becomes clear that no gene acts in isolation. The dream of genetic engineering—that we could precisely engineer organisms to have desirable traits without engendering any unexpected negative consequences—has proved to be a fantasy, as we discover that an organism might reconfigure itself entirely around an edit of just one gene. The part is inseparable from the whole.

Perhaps no field has been as crucial to this paradigm shift as ecology and Gaia theory. Ecosystems are rife with nonlinearity—symbiosis, positive and negative feedbacks, autocatalytic loops, trophic cascades, etc.—and homeostatic feedback mechanisms are how the planet maintains an environment hospitable to life.[7] Climate science recognizes this nonlinearity,

---

[7] At this point someone will hasten to unburden me of the notion that the biosphere maintains homeostasis; it is constantly changing, inviting newly fashionable terms like "homeodynamic." Well, a body is not always perfectly constant either, but both body and planet exhibit remarkable constancy over time in key regards. For example, the salinity of the ocean has remained almost the same for hundreds of millions of years, despite a constant influx of salts. Global temperature has remained constant within a range of a few percent, despite a large increase in solar irradiance. Oxygen levels in the atmosphere have likewise stayed in a range that allows animal life. And so on.

but it has failed to grasp its full implications, especially when faced with the necessity of translating its findings into the language of policy. Thus it is that it emphasizes global variables (primarily greenhouse gases) that we can, in principle, control with top-down strategies while deemphasizing the contribution of localized ecocide.

Through the dulling of our empathy and numbing of our feelings. First, a worldview that holds other beings as less sentient encourages us to think of them as mere beasts, mere vegetation, or mere dirt, undeserving of empathy. It contradicts our native heart intelligence and our pantheistic intuitions that understand we are in a world alive with sentience. The heart must therefore overcome the mind to convert its empathy into conscious action. Moreover, the repeated denial of our felt kinship with the world is a kind of trauma that quashes empathy. We repeat the trauma every time we denounce another person or ourselves as flaky, irrational, or mushy for espousing empathic rather than utilitarian motives for environmental action.

Second, empathy and the ability to feel are blunted by trauma. In the extreme case, severe childhood trauma leads to dissociation. It hurts so much to feel, that the unconscious mind in its wisdom provides a numbness, encysting the pain until the child grows up and becomes strong enough to process it. Until that healing can occur, the person will have a diminished capacity to feel. She may seem normal, but that is only because disengagement from feeling has become normalized in modern society. That is in part because of the elevation of reason and dispassionate objectivity in our culture, and in part because trauma itself has become normalized. I refer here to the less obvious forms of trauma that the horrifying prevalence of extreme physical, sexual, and emotional abuse of children, of war and political oppression, of domestic violence and economic poverty, can obscure. We hardly notice it, so normal has it become—as normal as the diminished capacity to feel that results from it.

Some amount of trauma is unavoidable in life and in fact necessary for development. Many traditional cultures recognized this when they included traumatic experiences as part of initiation processes—experiences that were held in a ritual container and integrated thereafter. In our society, it is often otherwise. Trauma is either kept a shameful secret, or it is hidden behind class, race, and gender stereotypes, or it is rendered entirely invisible through its normalization.

What society takes for normal is actually traumatic. To be yelled at by parents, to be shamed for one's sexuality, to be cast among strangers on the first day of school, to be exposed to intense depictions of violence on screen, to be deprived of frequent touch, to be confined to the indoors and playgrounds, to have ties repeatedly broken with frequent relocating, to experience the collapse of reality when parents divorce—if not all of these qualify as "trauma" in the same sense as outright physical abuse, they nonetheless contribute to a numbing of the capacity to feel.

On an airplane recently I watched a few minutes of an action movie that had been edited to make it more family friendly. Words like "bullshit" were changed to "bullshine" and "fuck" to "freak." Naked female breasts were edited out as well. Preserved, though, in its grisly entirety was a scene in which a man was fed headfirst through a giant meat grinder. Such images are both symptoms and agents of the normalization of trauma. As long as society agrees they are less disturbing than a woman's breast, can we ever hope to reverse the course of ecocide?

Finally, there is the trauma that accompanies economic, social, and political oppression. A brutalized or destitute person turns his or her attention to survival. It is not that the oppressed do not have "the luxury of empathy"—I have never found the poor to be less empathic than the privileged classes. If anything it is the opposite. However, the exigencies of daily survival might restrict the expression of that empathy to a narrow realm. Think of desperate displaced peasants in Brazil, working on roads, mines, or ranches in the Amazon. If they are numbed to the suffering of the forest, it is because they have to be in order to do what it takes to survive.

In general, survival anxiety is contrary to empathy, and it is not only the oppressed that suffer from it. The oppressors do as well. That is because we all live in a society of artificially created scarcity that hounds each of its members to get ahead of the rest. Our economic system has competition built into it. To make the most rational economic decision, often we must harden our hearts. Eventually the heart-hardening becomes a habit, a reflex, and a way of being.

Let me reconstruct the logic of the last few pages: in order to reverse the course of ecocide, we may have to consciously choose a healing path. We cannot count on collapse to compel us. In order to choose it, we need to change the conditions from which we are choosing. To change those conditions, we need to implement a different economic system and understanding of nature, and more importantly, we need to recover our empathic ability to feel. Therefore, the issue of environmental degradation and climate change cannot be separated from the need for social, economic, and personal healing.

To recover our ability to feel is going to hurt, because so much pain is out there waiting for us to feel it. It has been sequestered away, suppressed within ourselves and kept out of sight globally. On the outside, walls of cement and razor wire, walls of disinformation, walls of prisons, walls of gated communities, walls of historical blindness, and walls of complicit silence keep the dominant culture unconscious of the suffering of damaged peoples (human and otherwise). On the inside, it is false hopes, delusions, addictions, and pharmaceutical mind-control agents.

In the end, I do not believe that we can forever engineer our way out of the damage we have caused, any more than an alcoholic can forever postpone the pain by having another drink. Each technological response to conditions of increasing inhospitality to life involves greater and greater complication: more complicated social systems, more complicated technologies. Eventually, investments in complication reach a point of diminishing returns. Medicine, education, government, and the military are all groaning under the burden of bloated administrative structures that render them barely capable of carrying out the basic functions for which they were created. Eventually, such systems collapse under their own weight.

But again, the theoretical possibility of surviving on a ruined planet is a moot question when we could instead live on a beautiful, healed Earth. By extension, much of the debate about climate change is also superfluous for someone who is open to feel the pain to other beings that environmental damage causes. Whether or not it threatens our survival or that of our grandchildren, industrial civilization in its present form does grievous damage everywhere it operates. If this were the only form civilization could take, it might be an acceptable sacrifice.

I think a different kind of civilization is possible. The alternative and holistic, the indigenous and traditional, the innovative and inventive, and the regenerative and restorative show its contours. It is not just visionaries who have seen it. You, dear reader, have surely seen it too, bobbing in and out of sight as you struggle to keep your head above the choppy waters of habit and doubt. We are here to remind each other that it is there for the choosing.

8

Regeneration

Healing the Soil

In chapter 4 I wrote, "We are called to visit deep questions like 'What are we here for?' 'What is humanity's right role on earth?' 'What does the earth want?'" I wrote, "In the new relationship…, whenever we take from the earth, we seek to do so in a way that enriches the earth. We aren't unconscious of our impact, nor do we seek to minimize our impact. We seek to make a beautiful impact that serves all life."

This is the sponsoring precept of a growing movement that has adopted the adjective "regenerative" to describe its practices. The best known is regenerative agriculture.

Regenerative agriculture comprises an array of techniques that rebuild soil, water, and biodiversity. Typically, it uses cover crops and perennials so that bare soil is never exposed, fosters synergistic relationships among multiple food and nonfood crops, restores the natural water cycle, and grazes animals in ways that mimic herd animals in nature. In focusing on soil, it carries the original spirit of organic agriculture. The term "organic" was adopted by its preceptor, J. I. Rodale, in reference to the organic (carbon-containing) molecules that form living soil. He understood that soil is more than a mixture of chemicals. Unfortunately, the term has lost that original meaning, perverted to the point now where the USDA allows organic labeling on hydroponic vegetables that are grown in no soil at all. That is why I am not using the term "organic agriculture" here, although the regenerative practices I will describe are indeed organic in the true spirit of Rodale.

Regenerative practices have received attention recently for their ability to quickly sequester large amounts of carbon. As the reader knows by now, I think it is a mistake to evaluate technologies based on a single quantity like carbon, but in this case carbon corresponds to the building of topsoil. Topsoil is the foundation of life on land; it is the living layer of Earth's surface. Regenerative farmers and permaculture farmers understand that all beings on a farm, including human beings, will thrive when the soil thrives.

One promising technique to rebuild soil health is called management-intensive rotational grazing (MIRG), which seeks to raise animals in a way that mimics their role on natural grasslands. Unless you come from a culture that still practices traditional nomadic animal husbandry, when you think of pastured animals you probably picture an expanse of grass dotted with cows or sheep. That picture is far from anything you would see in a healthy ecosystem. Healthy ecosystems include predators for whom a field of scattered sheep would amount to an all-you-can-eat buffet. That is why herbivores gather in large herds for protection, intensively grazing an area and then moving on. That is what MIRG replicates.

As in natural grasslands, large concentrated herds of herbivores maintain grassland health and build soil. The herd eats mostly the sweet grass tops and tramples down and manures the rest of the plants before they can eat them down to the roots. This allows the grass to quickly recover after the herd moves on. Not only is the soil protected from erosion by the thick layer of trampled vegetation, but the damage causes the plants to send sugars into the roots, generating the rich exudates that, along with the manure and decaying plant matter, nourish soil biota. The soil biota, especially earthworms, increases the permeability of the soil, making it into a sponge for rain. The hooves of the animals aid that process by piercing the soil surface and making indentations that trap water.

When practiced on degraded land, MIRG brings it back to life. Dried-up springs begin flowing again, brown landscapes turn green, birds and biodiverse wildlife return, seasonal streams start flowing year-round, and depleted soils recover tilth and depth.

The most influential practitioner of management-intensive rotational grazing is Allan Savory, a Zimbabwean biologist and farmer who has inspired and taught his methods to farmers throughout sub-Saharan Africa, North America, South America, and Australia. His TED talk shows stunning before-and-after photos of land restored through the practice, which he calls Holistic Grazing.[1]

His claims have engendered considerable controversy.[2] I tend to believe the pro side: first, because the critics attack an uninformed caricature of the practices Savory promotes; second, because there has been a veritable groundswell of farmers and ranchers using the method and sharing anecdotes throughout the alternative farming media. However, actual proof is hard to obtain, in large part because of a paucity of hard quantitative data. In addition to the difficulty in measuring soil carbon, MIRG is not a standardized process, but must vary according to local conditions, even from one farm or valley to the next. That is the point in Savory's use of the term "holistic." The right practices can be determined only in intimate relationship to the land.

Although carbon sequestration data for MIRG practices is scarce, recent studies point to amounts that are much greater than most scientists previously believed. A 2014 study at the University of Georgia measured annual per-hectare increases in soil carbon at 8 tons per year on farms converted from row cropping to intensively managed grazing.[3] Water retention also rose by a third. The world uses about 3.5 billion hectares of land for pasture and fodder crops. Converting just a tenth of that land to

---

[1] Savory (2013).

[2] For a taste of the controversy, see Lovins (2014).

[3] Machmuller et al. (2015).

MIRG practices would (using the 8 tons figure above) sequester a quarter of present emissions. (MIRG also reduces methane emissions by as much as 22 percent compared to conventional meat production.)[4]

Individual farmers report much higher carbon sequestration figures. One of the most famous regenerative farms is Brown's Ranch in North Dakota, which used holistic grazing practices to raise carbon levels from 4 percent to 10 percent in the span of six years—equivalent to 20 tons of carbon per hectare per year.[5] Its rainwater absorption capacity also rose from a half inch an hour—entailing massive runoff—to eight inches per hour.[6] The rancher/farmer Gabe Brown and his family do not rely on managed grazing alone. They use a complex mix of cover crops, and multilayered intercropping as well: picture radishes and turnips growing under a canopy of sunflowers. They deliberately cultivate plants with a variety of root depths. The diverse, perennial vegetation nourishes high insect biodiversity, providing natural pest control—the farm has no problems with the corn root worm that plagues neighboring farms and is America's number one agricultural pest. Despite using no pesticides and no fertilizers, the farm yields 25 percent more corn than the county average, at a much lower cost per bushel.

Like grazing, regenerative horticulture also bears tremendous promise to draw down carbon, using similar principles. It avoids plowing or any other form of soil disturbance, favoring cover crops that are crimped or cut to feed soil biota and become the next layer of humus. According to research at the Rodale Institute, if instituted universally, organic regenerative techniques practiced on cultivated land could offset over 40 percent of global emissions, while practicing them on pastureland could offset 71 percent.[7] The potential for land-based CO_2 reduction is over 100 percent

---

[4]   DeRamus et al. (2003).

[5]   Hawken (2017), 73.

[6]   This figure and those following are from Ohlson (2014).

[7]   Rodale Institute (2014).

of current emissions—and that doesn't even include reforestation and afforestation.

Another impressive approach is Syntropic Agriculture, also known as regenerative analog forestry, developed in Brazil by Ernst Gotsch. In 1984 he purchased a huge 500-hectare farm that had been severely degraded due to clear-cutting, a manioc plantation on hillsides, and other abuse. It was known locally as "the dry land." Gotsch restored the land to health by mimicking ecological succession, by applying companion planting, and through heavy "chop-and-drop" pruning to build soil organic matter. Thirty years later, the land has been transformed. Fourteen dry springs have come back to life, streams flow all year round, the biodiversity of the original Atlantic coastal rainforest has returned, temperatures have cooled in the micro-region, and rainfall has increased. And the farm produces abundant food, lumber, and other products, including what some consider the world's highest quality cacao beans,[8] all without irrigation, pesticides, or fertilizers of any kind. As one worker there explains, food production rides on top of natural forest succession instead of fighting it. With each successive harvest, the soil is richer than the year before.[9] Projects inspired by his model have taken off throughout Brazil, and have spread to Australia and other places as well.

Gotsch did not develop this method with the goal of drawing down carbon, but according to a study by Cooperafloresta Brazil, his method sequesters about 10 tons/hectare of carbon depending on where it is in the successional cycle.[10] I mention this mainly to reiterate that the ecosystems-centered view of this book does not contradict the requirements of the standard climate narrative. Nor, however, does it depend on that narrative

---

[8] Taguchi (2016).

[9] See "Life in Syntropy," a short film about the farm and Syntropic Agriculture on Vimeo or YouTube.

[10] Cooperafloresta (2016), cited in Sendin (2016).

for its motivation—in the water frame and the biodiversity frame, regenerative practices are even more appealing.

Why Is Regenerative Agriculture Marginal?

Whether from the carbon frame, the water frame, or the biodiversity frame, regenerative agriculture makes ecological sense. It also makes sense from the point of view of food productivity. Why then, despite rising popularity, does it remain so marginal—agriculturally as well as scientifically?

The reason has to do with its incompatibility with ingrained ways of thinking, economic institutions, and scientific practices.

Farms such as Brown's and Gotsch's use a dynamic combination of many regenerative practices that are sensitive to unique and ever-changing local conditions. This makes it very difficult to isolate and quantify the effects of any single practice. A scientific demonstration of any of those practices would entail holding other variables constant across multiple test fields and control fields. This is not how regenerative agriculture works. One cannot apply standard processes to multiple parcels of land, because each place is unique. Therefore regenerative practices do not fit easily into current scientific protocol.[11] Furthermore, the pesticide, fertilizer, and genetically modified seed companies that fund most agricultural research have little incentive to fund studies of practices that require none of the above. Therefore, the data on the carbon sequestration, water retention, biodiversity benefits, and so on for regenerative agriculture remains mostly anecdotal.

The lack of hard data, in turn, prevents regenerative agriculture from entering a data-driven policy discourse. When environmental policy is based on quantitative greenhouse gas targets, how does one promote practices that do not easily generate quantifiable results? Compared to fossil

---

[11] The same point applies to holistic medicine. Because each body is unique, true holistic medicine is resistant to validation through controlling variables across standard diagnostic and therapeutic categories.

fuel emissions, underground biotic carbon storage is hard to measure even when we attempt it. Harder still to quantify would be the indirect benefits of biodiversity, groundwater recharge, and so forth.

It is not that policymakers find Gotsch's achievements unimpressive (his short film was shown at COP21). It is that they are hard to translate into current data-based policy. Ultimately we are being invited into a different way of engaging the world. Only dead things can be reduced to a set of data. A civilization that sees the world as alive will learn to bring other kinds of information into its choices.

Regenerative agriculture represents more than a shift of practices. It is also a shift in paradigm and in our basic relationship to nature.

Regenerative agriculture seeks to mimic and participate in nature, not dominate it. In the holistic view of regenerative agriculture, problems like low fertility, runoff, weeds, and pests are understood as symptoms of a disharmony between the farmer and the land. Instead of going to war against the problems, the farmer seeks to adjust her practices to restore soil health, water health, species distribution, and so on, in a process of deepening relationship.

In many respects, regenerative practices like organic no-till horticulture and management-intensive grazing fit poorly into the present agricultural-industrial complex, which favors standard products via standard processes with standard inputs at a predictable cost. Regenerative practices require intimate knowledge of the micro-conditions of each place. What works in Austria may not work in Cameroon, or even in the next valley over. What worked last year may not work this year.

There is no formula that says how long a herd should stay in each paddock for maximum soil regeneration. The farmer has to observe conditions and consider them in light of past learning, his own and possibly that of his father, his grandfather, his neighbors ... learning through trial and error and adding to the knowledge base. Likewise, there is no formula that says how deep to dig a swale and what to plant in it for best water retention. There is no formula for which cover crop mix is best. All of these are context-dependent. That means a farmer can never be a mere laborer.

For a regenerative agriculture system to work, farmers need to relate to land as to a unique individual. They must learn to listen to, see, and feel its needs and moods. Allan Savory goes without shoes to pick up subliminal information about the land he walks. The knowledge of a place builds over a lifetime and over generations, becoming embedded in a local culture. This kind of relationship is totally different from that of industrial agriculture, which treats parcels of land as just so many standard units, describable in terms of quantities of nitrogen, phosphorus, potassium, rainfall, pH, and so forth. Regenerative agriculture rejects the industrial model of production, with its pursuit of standardization and scale.

A food system that starts with intimate relationship to land will occupy a different place in society from the current system. For one thing, it requires a lot more time, and therefore a much larger segment of the population to practice farming and gardening.

At present, less than 1 percent of the population of the United States makes their living farming, down from two-thirds in 1850, half in 1880, and 10 percent as recently as 1955. Even in absolute terms, the farm population has fallen by 90 percent from its peak of about 32 million in 1910.[12] Other countries have showed a similar trend. Demographers and climate thinkers usually take its continuation for granted: one frequently reads statements like "By 2050, 70 percent of the world's population will live in cities."

This is a trend that must change if we are to live in right relationship to Earth. Urbanization is not some law of nature, nor an inevitable stage of human progress. Economic and technological conditions, among them the mechanization of agriculture and its conversion into commodity production, drive urbanization. Urbanization is uprooting; it is disconnection from the places of multigenerational cultural embedment; it is a distancing from the land. Yes, there is a role for cities; the archetype of the city will not and should not disappear from earth. Cities can be beautiful cauldrons of cultural ferment, alchemic crucibles that yield products only possible in an intense concentration of humanity. Yet for many, the land is calling;

---

12   Spielmaker (2018).

in fact, the trend toward urbanization is already showing signs of reversing in some developed countries. According to the 2014 USDA census, the number of young farmers began growing in 2007, reversing a longtime trend.

An objection to the call for more people involved in farming goes like this: "That's easy for you to say, because you are a privileged professional who doesn't know how hard and laborious farm work is." This objection fails on several fronts, not the least of which is that I do spend a lot of time doing manual labor on my brother's farm. Farm work in the context of industrial agriculture is very different from farm work on small, diverse, ecological farms. On the latter, tasks are diverse as well; rarely does one spend hour after hour, day after day, picking beans or driving a tractor back and forth. On the industrial farm, work is much like any industrial work: repetitive, routine, and dehumanizing. No wonder its final stage is its literal dehumanization, in the sense of replacing people with machines. The techno-utopian dream of transcending labor seems quite attractive if we take for granted what labor became with the rise of technology.

In the Story of Ascent, a move back to the land would be a regression. We were supposed to be progressing away from labor, away from materiality, away from the dirt. In that story, heavenly was better than earthly, high better than low, clean better than dirty, the mind better than the body, and the highest social classes the ones furthest removed from the land. Today's new computerized robotic hydroponic vegetable factories are its inevitable outgrowth. You can see how big a change it is to revere the farm again, and to reunite with the long-distanced material world.

Projections of growing urbanization take for granted the very thing that must change.

Feeding a Hungry Planet

I hope the above descriptions of regenerative farms put to rest the notion that industrial agriculture is needed to feed a hungry planet. Not only is it unsustainable long term, but it doesn't even outperform ecological agriculture in the short term. Again, quantitative proof of this claim is hard to come by. Most regenerative farms have no need to maximize food productivity per acre.

Some readers might protest that scientific studies typically show organic crop yields to be lower than conventional yields. Here we must look at what these studies take for granted. The high yields of small mixed farms are hard to measure because they typically produce multiple crops that may not find their way to commodity markets, but instead are consumed locally, via farm share programs or farmers markets, sometimes outside the money economy. Moreover, traditional forms of agriculture often employ multicropping and intercropping. So while an organic corn field will underperform a GMO corn field, what about the total yield of a corn field that also grows beans and squash, and is patrolled by free ranging chickens who eat the bugs? What about when insect-damaged fruit or vegetable seconds feed pigs or other livestock?

Optimal results come from long, even multigenerational, experience applied in intimate relationship to each farm. Comparisons of organic and conventional agriculture often use organic farms recently converted from conventional practices; rarely do they consider the most highly evolved farms where soil, knowledge, and practices have been rebuilt over decades.

I asked my brother, an organic vegetable farmer on 120 acres of land, what it would take for him to maximize food production in an ecologically sustainable way. (At present, he farms only about a tenth of that land.) In his typical laconic manner, he replied, "About two hundred people." If he converted the woodlands (severely degraded by 150 years of repeated logging) to agroforestry; if he put in water retention ponds and raised fish there; if he converted cultivated land to perennials and no-till intercropping; if he applied mob grazing to the pastureland; if he had a composting biogas operation to generate heat and electricity ... he could grow twenty times more food than he does today. But he doesn't have the two hundred people necessary to implement all these things; he has, depending on the season, between one and ten. So, his operation is based on generating a high output per unit of labor, not per unit of land.

This might explain why, around the world, small farms far outperform large farms in terms of yield. First observed by Nobel economist Amartya Sen in 1962, it has been confirmed by numerous studies in many countries. The best-known recent study looked at small farms in Turkey, which still has a strong base of traditional peasant agriculture.[13] Small farms there outproduced large farms by a factor of 20, despite (or because of?) their slower adoption of modern methods. Yet, the narrative of modern agriculture feeding the world is so strong that the OECD stated that "stopping land fragmentation" in Turkey "and consolidating the highly fragmented land is indispensable for raising agricultural productivity."[14]

Of course, small farms can be ecologically destructive just as large farms can, but in general, the worst abuses happen at industrial scale. Small farmers are much better able to put intensive care into their land, read its signs, and respond flexibly.

Figuratively and literally, we need to go back to the land. Unfortunately, U.S. policy has encouraged the opposite, aggressively pushing the interests of large agribusinesses around the globe. Thankfully, many countries, localities, and farmers have resisted this push. Most notably, France, Germany, Venezuela, and Russia have banned cultivation of genetically modified crops; Russia has banned their importation as well, as part of a nationwide transition to organic agriculture. This is about more than GMOs; it is about a whole model of industrial agriculture that goes along with it.

Is it practical to transition to a radically different model of agriculture in time to avert ecological catastrophe? My colleague Marie Goodwin attended a meeting of the Delaware Valley Regional Planning Commission about food security and arable land in the Philadelphia metropolitan region. The presiding official's presentation showed that the amount of farmland in the region was far less than that required to feed so many people should there be a breakdown in the global food system. Marie pointed out

---

[13] Ünal (2008).

[14] Monbiot (2008).

that if you included lawns, there would be enough land to feed everyone. The official was dismissive. "That's impossible," she said. "We could never get people to grow enough food at home to make any difference."

Marie points out that in the U.S. during the Second World War, victory gardens were responsible for 40 percent of all vegetables grown during that period: 9–10 million tons. It was an even greater percentage in Britain. This just goes to show how our notions of what is possible or realistic depend on cultural perceptions. Cultural perceptions can change, must change, and are changing. If by realistic we mean keeping everything the same, then we are going to have to stop being so "realistic."

Given the ruinous course we are on today, a more accurate word for "realistic" is actually "fatalistic." Again I'll quote Eileen Crist:

> In fatalistic thinking, the trajectory of industrial-consumer civilization appears set on tracks that humanity cannot desert without derailing; it is implied that while the specifics of the future may elude us, in broad outline it is (for better or for worse) a fixed direction of more of the same. Fatalism projects the course of human history (and concomitantly of natural history) as the inevitable unfolding of the momentum of present trends. By virtue of the inertia that massive forces display, from a fatalistic viewpoint, present patterns of global economic expansion, consumption increase, population growth, conversion and exploitation of the land, killing of wildlife, extinction of species, chemical contamination, depletion of oceans, and so on, will more or less keep unfolding.[15]

For there to be meaningful healing on this planet, "impossibilities" like more people growing food cannot remain impossible. We are indeed talking about a wholesale civilizational transformation.

Yes, it would require spending more time per capita on food production to feed ourselves and heal the land at the same time. It might require widespread home gardens, and government policies to encourage them. It might require 10 percent or 20 percent of the population to be involved in agriculture, not 1 percent. In a time of increasing global unemployment, this should not be a problem.

---

[15] Crist (2007), 54.

A model for a way forward might be found in Russia. In 2003, Russia promulgated the Private Garden Plot Act, which entitled every citizen to a tax-free private plot of several acres of land for gardening or recreation and accelerated the dacha and ecovillage movement there. As of 2016, small plots provided nearly half of Russia's food.[16] In many developed countries, though, agricultural regulations, zoning laws, building codes, and so on make it difficult if not illegal to farm ecologically, especially for the small farmer. In America, for example, concerns about food safety have led to prohibitions on mixing livestock and crops. No more ducks eating slugs or chickens controlling insects. No more dogs protecting fields from woodchucks and deer. Complicated regulations that were created to rein in large producers' irresponsible behavior can be prohibitively time-consuming and expensive for the small farmer, who does not have a compliance department to manage the paperwork. The regulations were created for, and to a large extent by, large producers. Newly proposed regulations require documentation of each time livestock is moved. This is no problem for a confinement operation with thousands of hogs or chickens that are occasionally moved en masse. It is impossible for a small ecological farm to comply, when it may have a few dozen head of livestock and a small flock of poultry that is constantly moving.

Outside agriculture, other regulations are misaligned with ecological needs. Tiny homes do not meet building code size requirements. Homes using composting toilets and graywater aquaponics systems nonetheless have to install expensive and unnecessary septic systems.

To align our society with ecological healing is not impractical. It just needs a shift in our perceptions, priorities, and laws. Nature's tendency is toward wholeness, if we only align ourselves with it instead of fighting to keep things the same.

On the level of national and global policy, a transition to regenerative agriculture would require significant political will and leadership. Many farmers today are stretched to the limit by their debt obligations, making it

---

[16] Russian Federation (2018).

impossible to afford a few years of lower income while transitioning their farm. Some kind of public subsidy is needed to support the transition. I think the best way to accomplish this would be to reassign existing subsidies (agriculture is already highly subsidized in many countries). In the U.S., some 85 percent of farm subsidies go to the largest 15 percent of farm operations.[17] Annual farm subsidies are at least $20 billion in the U.S. and even more in the EU. Using just half that, one hundred thousand American small farms a year could each receive a $100,000 three-year transition subsidy. That is a slow enough pace to avoid disruptions in the food supply, and fast enough to make a significant ecological difference. I will be happy to donate the dinner napkin on which I made these calculations to Congress.

Then there is the labor problem—except that there isn't one. Here again we can simply redirect existing resources. Youth unemployment is at least 10 percent in the U.S. and nearly 20 percent in Europe. Moreover, governments around the world, especially the U.S., spend vast sums to induce young people to join the military, or even require them to do so. Among the American working class and underclass, many choose military service because of an idealistic desire to serve, coupled with a lack of economic opportunity in any other field except illegal drugs. Unfortunately, this idealism depends on obsolescing narratives such as "America, bringing liberty and democracy to the world" that are actually cover stories for imperialism. As the age of empire dwindles, these narratives are losing their power, yielding to a creeping cynicism within the military, and especially among veterans. If I may make an immodest proposal, what if we met these twin needs of service to the world and economic security by creating an eco-corps, dedicated to ecological healing and to the service of all life on earth?

Ecological healing works in both directions: working with plants, animals, soil, and water has powerful therapeutic benefits.[18] Modalities like

---

[17] Smith (2016).

[18] In case you need peer-reviewed confirmation of this assertion, which seems obvious at face value, you can start with: Soga et al. (2017).

horticulture therapy and garden therapy show impressive results for at-risk youth, prisoners, veterans, and people with chronic diseases, not surprising when we understand health as wholeness, and disconnection as disease. Psychiatric conditions in particular improve with interaction with nature, lending credence to the view that most of them are symptoms of "nature deficit disorder." Conditions like ADHD, depression, and anxiety often improve or disappear entirely when the individual interacts regularly and meaningfully with the natural world. The healing of individuals, society, and the world go hand in hand.

Healing the Water

In the carbon-centric climate narrative, ecological healing efforts that do not directly sequester carbon receive little attention. That must change if, as I have argued, water is as important or more important than carbon in maintaining climate equilibrium.

To be sure, all of the regenerative agriculture practices I've described bear huge benefits for the water cycle. In contrast to Brown's Ranch with its eight-inch-per-hour absorptive capacity, on conventionally farmed land most of the water from heavy rains either runs off (carrying topsoil with it) or forms puddles on the ground that quickly evaporate. Then the water never recharges the aquifers.

As a general rule, what is good for the soil is good for the water. Hydrological health is a happy side effect of soil health.

There are also practices designed with the explicit purpose of restoring water health, for which building soil may be the happy side effect. These practices are especially significant in desertifying areas, where they are stopping and even reversing the process of desertification.

India is among the countries most vulnerable to water scarcity. Heavy use of groundwater for irrigation has led to plummeting water tables and dried-up wells. The solution, typically, is to dig deeper wells—a patently myopic response. But in Rajasthan, Rajendra Singh, known popularly as the "waterman of India," has inspired a movement to implement low-tech water retention structures, reviving a technology thousands of years old. These structures include *johads,* which store water for future use in a way that also allows it to slowly permeate into the ground, earthen dams to create small reservoirs, and check dams to slow runoff after heavy rains and allow greater penetration into the water table. His work ignites a "virtuous circle": greater water availability leading to more vegetation, leading to less soil erosion, leading to more water penetration, leading to more groundwater availability. Farming also becomes more productive, so that like the soil, the rural population no longer runs off to the cities. More local labor means more capacity to maintain the *johads* and dams. Singh's ideas have been implemented in over a thousand villages, which maintain thousands of water retention projects and have planted millions of trees. Five dormant rivers in the region have come back to life and flow year-round.[19]

A related concept to Singh's work is the "water retention landscape," which uses terraces, berms, swales, and ponds to catch water during the rainy season so that it soaks into the water table instead of running off. My first encounter with water retention landscapes was at the Tamera Ecovillage in southern Portugal, a desertifying region where streams that once flowed year-round now flow only seasonally. Arriving at Tamera after several hours of driving through dusty, brown landscapes, I was stunned by an explosion of green. Fruit trees, gardens, and forests surrounded several ponds and small lakes—this in the middle of a drought-stricken summer. My first suspicion was that they had been pumping out groundwater. No, it had accumulated from last winter's rains, held in the lakes by earthen dams. My second suspicion was that this community (whose majority are German expatriates) had arrogantly engineered and dominated the landscape. "How did you decide where to put the lakes?" I asked. They replied that they closely observed the land for several years until they understood where the water wanted to be. This attitude exemplifies the intimate relationship with the natural world that must come prior to any system of methods or practices.

---

[19] Ahmed (2015).

Water retention slows the process of rain returning to the sea, so that it completes what the brilliant and controversial Austrian scientist Viktor Schauberger called the "full water cycle." In the half water cycle, water evaporates from the ocean, falls as rain over land, and flows into streams and eventually back to the ocean. In the full cycle, the rain sinks into the earth and spends anywhere from weeks to decades there before emerging from springs. Working in the early twentieth century, Schauberger was an early critic of deforestation, which he recognized as a culprit in truncating the full water cycle.

Water retention can be accomplished in urban areas as well, through permeable surfaces, tree planting, catchment basins, and household rainwater storage cisterns. Without such measures, cities can do tremendous damage to surrounding communities and ecosystems. Los Angeles sparked the infamous Water Wars of Southern California in 1913 when it began diverting the Sierra Nevada snowmelt, and continues to dominate the region's water resources, spending a billion dollars a year to source water. Meanwhile, the city also spends half a billion dollars getting rid of water—through its stormwater system. Like many places, LA has both too much water and too little water, flood and drought. One naturally follows the other: floods and droughts are both consequences of low rainfall absorption. Water retention, also called "slow water," can ameliorate both, turning both deserts and cities green again.

What is possible on ranches and cities is also possible on a gigantic scale. One of the most impressive desertification reversal projects is the Loess Plateau Watershed Rehabilitation Project in northern China, made famous by filmmaker John D. Liu. The loess plateau of this region was a cradle of civilization, exceeded only by Mesopotamia in its antiquity, and suffering a similar fate. According to Liu, by the year 1000 deforestation and unsustainable agricultural practices had reduced a lush land of forests and grasslands to a parched, eroded wasteland that looks like a desert even though it receives modest rainfall. That is because 95 percent of it runs off immediately, forming huge erosion gullies and giving the Yellow River its characteristic color.

The results of the rehabilitation project can be seen in Liu's stunning before-and-after photographs—the land has literally come back to life. The change came through a huge investment of labor, money, and planning. Local residents were recruited in large numbers to build small earthen dams, terraces, and other water retention features. They planted trees, abandoned slopelands unsuitable for planting, and restricted the grazing of sheep and goats. Crucially, they closely participated in the planning of the project as well, and were given subsidies for their labor and granted land rights to restored areas. In the end, an area of 15,600 square kilometers (the size of Belgium) was restored at a cost of about half a billion dollars. The written word cannot do justice to the changes that transformed the landscape, but Liu's film has inspired similar projects in Rwanda, Ethiopia, Jordan, and other countries.

These projects show what is possible when the collective human will is brought into alignment with Earth's capacity to heal. It shows what is possible when we make a collective choice toward beauty rather than quantity. And here is the caveat—it does require will, an active choice. Otherwise, we will continue to slide in the direction of our inertia.

Is something like this possible globally? Is it practical? Realistic? No, if we accept the permanence of society as we know it. Yes, if we are prepared to let go of what had seemed unchangeable. A half billion dollars over ten years is nothing in comparison with, say, global military budgets, which total about 3,300 times that. Devoting just 10 percent of military spending to watershed restoration would fund 330 loess-sized projects. Earth is actually not asking very much of us.

All right, I admit it, that last sentence is a bit disingenuous. Earth is asking a lot of us. Earth is demanding a transformation of our civilization's fundamental priorities. Earth is demanding that we see her as sacred. Earth is demanding that we see her as alive. Earth is demanding we reorder our civilization and all its institutions accordingly. Money, government, law, technology ... all must change. That is why the ecological crisis is truly an initiation for humanity.

The Mutual Need of People and Planet

When I speak publicly about the planetary crisis, often someone in the audience will tell me, in remonstrative tones, that it isn't a planetary crisis at all. The planet will be just fine regardless of what we humans do. The threat isn't to Gaia; it is to humanity.

This well-worn assertion appears to embody humility with respect to the vast power of nature; in fact, it represents a subtle form of human exceptionalism and a disregard for nature's purposiveness. If we affirm that Gaia is a living being with a life cycle and a destiny, then we can only assume that humanity was born for an evolutionary purpose. Each species, each child of Gaia, has a role to play, and we are no exception. The fulfillment of that role is thus of crucial planetary importance.

Imagine saying of a mother with a gravely ill child, "The threat isn't to the mother or the family; it is to the child. Don't worry about the mother. If the child dies she'll be fine." Only if we understand life as a happenstantial film of biochemistry on an orbiting rock, can we ignore the intuition that humanity was entrusted with gifts and bound by love to serve the evolution of planet Earth. Only by denying that Gaia is a coherent, conscious, purposive being can we imagine that humanity's survival does not matter.

Nature does not produce a new species by accident. Ten or twenty years ago, this statement would have seemed blatantly unscientific, contradicting as it does the principle that evolution happens only through random mutation followed by natural selection; but today the study of epigenetics and biological genetic engineering makes it clear that genes, organisms, and the environment evolve together in a tightly coupled nonlinear partnership. Evolution is purposeful.[20] No, I am not advocating Intelligent

---

[20] Thirteen years ago when I first began telling people I was a Lamarckian, I was met with eye rolls or blank stares. But last week I confessed it to a biologist I met at a conference and he didn't bat an eye. "Everyone is a Lamarckian now," he said. "Lamarck was right." This is no longer fringe science. I refer the interested or skeptical reader to James Shapiro's *Evolution: A View from the 21st Century,* Denis Noble's *Dance to the Tune of Life,* and Scott Turner's *Purpose and Desire.*

Design, unless it is the intelligence inherent in nature itself. Nature itself has purpose—it doesn't need a divinity to impose it from the outside. That God, made in the image of the human engineer, is retiring. The new God doesn't impose intelligence onto a lifeless mechanism of a universe. The new God *is* the intelligence of a living, sacred universe. The purpose that guides the evolution of species comes from larger, living wholes. The environment creates organisms for its purposes, as much as organisms alter the environment for theirs. The parts create the whole, and the whole creates the parts.

The Whole has created humans too for its purpose.

There is a certain comfort in thinking that the planet will be fine without us, yet there is also a certain fatalism. It is akin to the fatalism that comes in response to disconnection from one's destiny. It induces a kind of aimlessness. As humanity exits the old Story of Ascent and its triumphant techno-utopian destiny, we are indeed experiencing a collective aimlessness. In that story, our purpose was ourselves. That purpose has been exhausted. We are ready to devote ourselves to something greater.

In the Story of Interbeing, entrusted with gifts and bound by love, we realize that our passage through the present initiatory crisis is of planetary moment. Out of the wreckage of what we thought we knew, something else may be born.

Tending the Wild

The regenerative practices I've described are rooted in a mindset and way of relating that goes back tens of thousands of years outside civilization, and even as a recessive gene within civilization, the seed of the future.

This section is named after a book by Kat Anderson that describes the relationship between the pre-colonial indigenous people of California and the land. Anderson demolishes the myth that hunter-gatherer people were mere occupants of pristine "nature," demonstrating their deliberate, sustained influence on the composition of biotopes and species in their territory. Entire landscapes that appeared to the untrained eye of white settlers as wild were anything but. Anderson explains:

> *Through coppicing, pruning, harrowing, sowing, weeding, burning, digging, thinning, and selective harvesting, they encouraged desired characteristics of individual plants, increased populations of useful plants, and altered the structures and compositions of plant communities. Regular burning of many types of vegetation across the state created better habitat for game, eliminated brush, minimized potential for catastrophic fires, and encouraged diversity of food crops. These harvest and management practices, on the whole, allowed for sustainable harvest of plants over centuries and possibly thousands of years.*[21]

When white settlers marveled at the stupendous bounty of fish, game, and wild plant foods that the Indians, it seemed, lazily lived off in an indolent existence, when John Muir wrote his glorious praise of California's Central Valley with its endless meadows of wildflowers, they were actually looking at a sophisticated garden, lovingly tended for generations. According to the elders Anderson interviewed, "wilderness" was not a positive concept in Native culture; it meant land that was not well tended, land in which human beings were not exercising their duty to protect, enhance, and develop life.

It is easy to see how the perception of indolent natives living in "virgin" wilderness facilitated intrusion onto their territory. After all, they were just inhabiting it; they weren't developing it, they weren't doing anything with it. It was going to waste. The ideology of the wild is of a piece with the ideology of conquest.

What looked to European settlers like untamed wilderness was actually the product of millennia of intentional human influence. Calling it wilderness, or "virgin territory," gave them license to occupy it, cultivate it, "develop" it, and "improve" it.

This attitude is still wreaking its damage today in places like Brazil, where Amazonian tribes attempting to establish rights to their ancestral territory are required to prove the traditional occupancy of that territory.

---

[21] Anderson (2006).

The difficulty is that their mark on the land is not of a kind that the government can easily recognize. They did not establish permanent farms or dwellings. The ideology of "the wild" renders the mutuality of their relationship to the land invisible.

Modern conservationists might be excused for wanting to minimize human impact, since the kind of human impact we've seen in the industrial era makes the caring observer recoil in horror. We might be excused for promoting an ethic of "leave no trace." We might be excused for envisioning a future where humanity retreats to bubble cities, space colonies, or a virtual reality, leaving nature behind to recover its former wholeness, relating to it as a spectacle or a venue for recreation, visiting it perhaps as zero-impact ghosts, observers but not participants.

Tending the Wild suggests a different vision, freeing us from the perceptions with which industrial society has imbued us. Instead of zero impact, it suggests positive impact. Instead of leave no trace, it suggests "leave a beautiful trace" or "leave a healing trace." It suggests that we ask, "What is our proper role and function in service to the health, harmony, and evolution of this whole of which we are a part?"

We have potent gifts of hand and mind that take the form of technology and culture. These gifts are not meant for us alone. They are meant to serve the wholeness and evolution of Life.

True, as a civilization we have not used our gifts in this way. Even pre-industrial peoples wreaked considerable havoc, contributing to the desertification of the Middle East and other areas, and to the disappearance of North and South America's megafauna. The latter coincided with dramatic changes in the composition of vegetation: the extinction of mammoths, mastodons, and other megafauna led to forests replacing savannas in many parts of the continent, along with a steep decline in overall biodiversity[22] and nutrient availability.[23] Perhaps it was through the tragedy of these

---

[22] Barnosky et al. (2016).

[23] Doughty et al. (2013).

extinctions that the newcomers to North America learned to pay special attention to preserving and expanding what remained of the continent's biological wealth.

Just because someone is indigenous does not mean he or she, or her culture, knows how to live in mutually beneficial harmony with the earth. It is something each culture must learn. Furthermore, each level of developmental scale requires a new learning.

Extinctions of megafauna and other animals and plants regularly followed human settlement of new lands. Australia, the Americas, New Zealand, Madagascar, and Polynesia all experienced them, suggesting a kind of inevitability to anthropogenic ecocide, which has only accelerated along with our capacity to perpetrate it. Yet, in the end, people in all of these places eventually came into equilibrium with their lands. In most places, as the subsequent biological wealth of the Americas exemplifies, it was an abundant and biodiverse equilibrium. This suggests another possibility beyond Man the Destroyer—that we can learn from our mistakes, that we can mature in our gifts and turn them toward a different purpose.

If so, then we have a lot to learn from indigenous people who sustainably tended and enriched the lands and waters they called home. Sometimes this might entail learning from their actual methods, but more probably it is a matter of adopting the mentality that gave birth to those methods in the first place, since the environment of ten thousand or even five hundred years ago is probably lost forever. That mentality is the product of the worldview I call the Story of Interbeing, a story that unites the diverse mythologies of indigenous peoples. More practically, it means forging intimate, respectful relationships with nature in its specific, local embodiment. Through extended close observation and interaction with nature, we can begin to hear answers to questions like "What does the river need?" "What does the mountain want?" "What is the dream of the land?"

These are the kind of questions that may take long, intimate observation—scientific and otherwise—to answer. There is no sure formula to determine which land should be grazed with compact herds, and which land should be protected from grazing entirely. There is no sure formula to determine which exotic species should be controlled as an "invasive," and which should be welcomed for their contribution to a new balance.

This latter question is reflected in the debate between restoration ecologists, who seek to reverse ecological damage and bring back native species, and the "new ecologists" who question the assumptions behind the concept of restoration. Science writer Janet Marinelli describes the divide:

> At a time when restoration of forests and other ecosystems is increasingly essential, the dominant paradigm of restoration science has been shaken to its core. Restoration ecologists, for whom returning lands to their state before the arrival of Europeans on the continent is still the basic, if rarely stated, goal, have been at loggerheads with so-called new ecologists, who challenge the primacy of native species in conservation thinking and champion the "novel ecosystems" of native and exotic species that increasingly dominate the planet.[24]

This passage hints at an emerging synthesis between the two positions. Human intervention is obviously necessary to restore ecosystems—not necessarily to their former state, but to a state of health. Yet, neither is their former state an irrelevancy. Historical knowledge is useful in understanding the land's needs, where they came from, and how to meet them. No general formula can tell us, for example, when an invasive species needs to be controlled, and when it is actually an agent of a damaged ecosystem coming back into balance.[25]

Neither "trusting nature" nor "restoring ecosystems" offers a reliable recipe for action. The question is not whether to participate, but how. Absent a recipe, we are left with place-specific, intimate observation and sincere inquiry informed by the understanding of the nonlinear, living nature of each ecological being. That's how we gain the wisdom to know

---

[24] Marinelli (2017).

[25] For a critical discussion of the complexities of invasive species management, see Tao Orion's book *Beyond the War on Invasive Species*. Often the efforts to control invasive species do more harm than good.

how to participate in regenerating the health of the places and planet we inhabit.

All of the regenerative practices I've described in this chapter partake in a common sponsoring idea. Earth is alive. What is alive, we can love. What we love, we wish to serve. When what we love is sick, we want to ease its suffering and serve its healing. The more deeply we know it, the better we can join its healing.

9

Energy, Population, and Development

The Paradigm of Force

As climate change has become the number one focus of environmentalism, discussions about sustainability are increasingly conversations about energy. Unlike biodiversity or ecosystem health, energy is easy to measure, tempting the quantitative mind to equate a sustainable society with sustainable energy sources. Thanks to energy's amenability to quantitative analysis, we try to extract the issue of energy from a matrix of social and ecological dependencies.

When I first took on writing a chapter about energy, my own quantitative mind leapt at the opportunity to delve into nice clean numbers. I got wrapped up in the minutiae of emissions intensity per kWe, "energy return on energy invested" (EROEI), the relative benefits and risks of various kinds of renewable energy, high-end and low-end projections, and so forth. I felt duty-bound to weigh in, unconsciously influenced by the idea that any "serious" discussion of the energy issue must engage quantitative reasoning. I wanted to know, is it possible to transition to sustainable energy, or isn't it? The more I read, the more unclear it became, as various authorities stand in foursquare contradiction to each other. I became distracted and depressed. Let me check my email again instead. How about reading some more articles. Or maybe I'll watch *Game of Thrones.* I began to understand the seeming apathy and passivity of the public. How different that feeling was from the state I'm in when I'm doing my best work: I can hardly keep away from it. Laziness, resistance, procrastination ... maybe these are not problems; maybe they are symptoms, maybe they are the voice speaking to the man in the maze, saying, "Just stop."

So I did. And I understood that in all I was reading, the questions weren't big enough and the best-case scenarios weren't good enough. We can have a paradise on earth, if we only wake up to that choice. The ecological crisis is supposed to be the wake-up call, not a challenge to overcome to stay on our current course.

That is why this book will spare you the graphs and data proving that renewable energy offers a viable future. Or that it doesn't offer a viable future. Metrics such as EROEI seem at first glance to offer a clean, simple way to evaluate various energy sources, but these seemingly objective numbers typically involve a lot of assumptions and projections that are the grist for endless debate. EROEI tends to be lower in practice than in theory; then there is the matter of accounting for grid dependency. Should the numbers for photovoltaics include some portion of the fossil fuel capacity needed (at the present juncture) to back them up? How do we account for efficiencies of scale and technological improvements? How will other technologies coevolve with energy technology? How will social patterns change? We face the same problem here as we do in removing a species from its web of ecological relationships for carbon accounting. When dealing with a complex interconnected system, there is no such thing as an unbiased number. That is perhaps why published EROEI figures for solar vary so widely, from as low as 0.83 (making it an energy sink)[1] to as high as 14.4.[2] The internet

---

[1]   Ferroni and Hopkirk (2016), 336–44.

[2]   Koppelaar (2017).

abounds with authoritative demonstrations of why a transition to a renewable energy future is inevitable, and equally authoritative demonstrations of why it is impossible. When I read either of these two positions, I feel stupid for ever having believed the other.

As with the whole climate debate, the energy debate distracts attention from more fundamental issues. What is most significant is that which both sides accept without question. The energy debate takes for granted that it would benefit humanity to continue using a lot of energy (provided it can be done sustainably). It takes for granted current conceptions of development. It takes for granted that human well-being has progressed thanks to increasing energy consumption, which has freed us from toil and allowed each person to benefit from the equivalent of the labor power of thousands of people. It takes for granted the beneficence of systems of medicine and agriculture that require large energy input. In essence, it takes for granted the desirability and necessity of "progress" as we have known it, associating progress with a growing capacity to impose our will on the material world.

That does not mean energy is a nonissue. Something else is at play in our obsession with energy, beyond its easy fit into the quantitative mindset. The harnessing of nonfood energy sources to perform work is virtually unique to human beings. For half a million years, human beings have deliberately used fire to transform materials; for five thousand years, we have used animals to plow fields and carry loads; for several hundred years we have burned coal, oil, and gas to power industrial technology. Our profound impact on the planet owes itself to our harnessing of energy to do work.

What is unique to human beings is also defining of human beings. Who we are as a species relates intimately to what we use for energy. The old story of man over nature extrapolated the exponential increase in energy use into the future, assuming that nuclear power would be as great a leap past fossil fuels as fossil fuels were past firewood and oxen.

The dreams of the early atomic scientists proved vain. Energy availability per capita did not continue its post–World War II exponential rise, but leveled off and in many places even began to decline, nuclear power or no.

Energy consumption per capita peaked in America in the 1970s; worldwide it continued to rise as industrialization brought the old technologies of coal and oil to the rest of the world, but it did not rise exponentially. We are nowhere close to replicating the twentieth century's tenfold increase in energy consumption. "Futuristic" technologies like nuclear fusion are unlikely to reignite the exponential growth that the fossil fuel revolution taught us was normal: (hot) nuclear fusion has been "a few decades away" for my entire lifetime.

What appears to us to be futuristic depends on our conception of the future, which reflects present-day conditions and thinking much more than the actual future. If human progress means progress in dominating nature, then it depends on ever-increasing energy sources. In a cosmological system that denies innate intelligence to matter, an ordered world depends on human ability to impose that order, to move and transform the building blocks of matter. The more energy we have at our disposal, the greater the scale upon which we can impose order.

The Story of Ascent narrates human history as an ever-ascending capacity to impose order onto chaos, intelligence onto randomness, and civilization onto the wild. This triumphalist narrative is breaking down in our time, as the promised technological paradise has receded into the future, a future we now suspect will never come. Instead, conditions on the planet are worsening, to the point where many fear for the survival of civilization entirely. "The conquest of nature" no longer bears the exciting cachet it once did; many of us now reject the very idea of it. Bubble cities and robot servants no longer populate our visions of paradise; we aspire instead to Edenic harmony between human and nature.

Shifting from domination to participation, we understand that we can improve life by cooperating with natural processes, not overcoming them. It takes a lot of energy to run a system of industrial agriculture requiring a war on weeds, bugs, and fungi, and the constant domination of soil chemistry. It takes a lot of energy to run a high-tech medical system based on killing germs and dominating body processes. A force-based system requires a lot of energy—that's a basic principle of physics. While this approach is extraordinarily effective in many situations such as acute trauma, its energy-intensive, money-intensive methods are far inferior to holistic practices when it comes to most chronic conditions.[3]

And what qualifies a method as "holistic"? It is one that draws on an understanding of the interconnectedness of all things, the intimate connections between self and other. It recognizes an all-pervading intelligence with which we can enter alliance, manifest as the intelligence of the body, the soil, the forest, the ocean, or the planet.

I am not a researcher in the field of holistic medicine, but my personal and secondhand experiences have convinced me that inexpensive, natural methods can cure most "incurable" medical conditions. Intuition tells me that what is true for the human body must also be true for the ecological body, the social body, and the body politic. None of them will be healed by finally mustering enough control to eliminate weeds, terrorists, violence, germs, and so on once and for all. Only by allying ourselves with innate tendencies toward wholeness is a healed world possible.

At risk of drawing too neat an analogy, what most alternative treatments have in common is a respect for the natural healing capacities of the human body, supporting and aligning with them rather than seeking to dominate and control them. They are not based on force. What miracles might be possible if we serve the regeneration and wholeness of Gaia? What might be possible if we strengthen her organs, detoxify her tissues, unblock her fluids?

The Meaning of Development

It takes energy to exert force; therefore, the energy crisis invites us into a less forceful way of life. To equate that with a regress in human well-being reflects the prejudices of our time. Another mode of development beckons.

---

[3] I'm not going to cite a reference here—if you are skeptical you can do your own research. In some areas like hypertension and heart disease, the evidence for the benefits of such things as yoga and meditation is abundant even for those who will only accept quantitative peer-reviewed studies.

Holistic medicine is not more energy-intensive than high-tech medicine; it is less. Regenerative agriculture is not more energy-intensive than chemical-dependent agriculture; it is less. Close knit communities and extended families are not more energy-intensive than the single-family home; they are less. What's more, these alternatives produce better health, abundant food, and more happiness than the present models.

Therefore, the main question in sustainability discourse is the wrong question. "How do we meet humanity's growing demand for energy?" carries assumptions that needn't be true. We can shift to a mode of development different from the Western model of the last several centuries.

A woman in India told me several years ago, "I grew up in a household of more than a hundred people. Aunts, uncles, cousins ... several generations were under one roof in our compound. As children we always had lots of people to play with, and every adult we saw was someone who loved us."

"But then," she said, "it all changed. The family got rich, and now everyone moves out of the compound when they get married to form their own household. No one stays in the village. We are all much wealthier now, each owning our own home and our own car, but no one is as happy as we were when I was a child. When couples argue there isn't anyone to hear them. There is no one to help look after the children and no one for them to play with."

Something similar has happened around the world, as small towns and villages have emptied in a tide of urbanization propelled by the ideology and economics of development. Many people assume such development is inevitable and desirable: that the manifest destiny of every Chinese peasant and Indian villager is to arrive at an American lifestyle, driving around in a private automobile, living in a three-thousand-square-foot house with separate bedrooms for each child, shopping at a supermarket for food grown a thousand miles away, entertaining themselves with digital media. Maybe, the optimists think, it could work if their electric grid is powered by wind turbines and their cars are fueled by ethanol or hydrogen or run on electricity. Meanwhile, pessimists point out various difficulties in achieving such a lifestyle for all without overstepping various planetary boundaries.

Lost in the debate is the question of whether such an outcome is desirable. More and more people in the West are now realizing it is not. Many who achieve the American Dream—a dwindling prospect—discover it to be the American Nightmare. I live in a country where nearly one in five people takes psychiatric medication for depression and anxiety, where suicide and addiction are at historic levels, where a third of all children suffer abuse, where half of marriages end in divorce. These afflictions transcend race and class. Neither privilege nor success is proof against them.

Realizing the bankruptcy of the Dream, many people in the West walk out of it. Some do it consciously and others not (the paralysis of severe depression or the escapism of addiction might be considered a sort of unconscious walkout). Those who do it consciously seek to live in a different way. They get their hands in the soil. They live communally. They unplug from digital media. They downsize their houses and their incomes. They seek to learn from cultures who have preserved other ways of living. Today they exist mostly on the margins of society, but it needn't be so. They embody an invitation into a once and future way of being.

Rather than think about how to sustain an energy-intensive social infrastructure, we should think about transforming it altogether—and not for the reason of using less energy. A shift in values toward the local, the participatory, the embodied, the communal, toward wholeness and empathy, toward the restoration of ecological relationships, will necessarily reduce energy consumption as a side effect. That is inevitable, when we no longer try to dominate and control the world through force. It is inevitable when we accept ourselves as a part of it and listen for how we might participate in its unfolding.

The reason to deurbanize, relocalize, downsize, re-skill, return to the soil, and live in community need not be to reduce energy consumption or cut greenhouse gas emissions. These and other quantifiable benefits that result are barometers of health and not its essence. The reason can be to restore the connections that make us happy, to come back into relationship with each other and with the beings of nature, to live in a way aligned with the Story of Interbeing, which says that relationship is *who we are.*

There are many commentators in the Post-carbon and Transition Town movements who well understand the futility of trying to switch to renewable energy sources while keeping everything else the same. Most, though, still subordinate social transition to the exigencies of climate crisis or energy shortage. Either they say, "We cannot maintain our standard of living as it has been, and will have to prepare for a downgrade," or they say, "Let us begin doing the things that are necessary to transition to a post-carbon world." They subordinate choice to necessity. What if, as suggested in chapter 7, we will not be forced into an ecological future? What if we must choose it?

Many are already choosing it, as best they can within an existing social and economic matrix that is hostile to it. I don't hear anyone saying, "I left corporate America and took a permaculture design course because I had no choice." It may very well be that a personal crisis created the conditions for that choosing. It may be that the choice to value a beautiful life over a secure life was until that moment unavailable. We didn't know that it was an option, until some kind of collapse cleared away an old story. We are not the originators of our choices. We are but the choosers.

The multiple crises converging upon us today present us with new opportunities for choice. We face not a downgrade but an upgrade in our quality of life. Our "standard of living" may decline: standards are a function of quantitative measures. The choice to live in greater wholeness is also a choice for smaller quantities of things like indoor floor space per capita, BTUs of energy consumed, total miles traveled, dollars of health care purchased, volume of global trade, hours of paid professional day care, quantity of prepared foods purchased, total annual timber harvest, and so forth. But unless you think that people are happier in their McMansions than they were in the small homes of yesteryear, unless you think that children are happier being carted around to organized activities than they were playing freely outdoors with a pack of other kids, unless you'd rather fly than walk to visit your dearest friends, our choice can represent an active "yes" to a more beautiful world, not a capitulation to grim necessity.

One way to conceive of the choice before us is that it is to shift from quantitative to qualitative values. To quantify is to master; it is to reduce the infinite variation of the world into standard units. It is to make the world ours, to order it according to our measures. This conceptual imprisonment of the world sets the stage for its actual imprisonment. Unfortunately, as in all prison societies, the jailor becomes a prisoner too. Thus we have become stuck in the endless pursuit of more, more, more.

In *The More Beautiful World Our Hearts Know Is Possible,* I wrote,

> *How much of the ugly does it take to substitute for a lack of the beautiful? How many adventure films does it take to compensate for a lack of adventure? How many superhero movies must one watch, to compensate for the atrophied expression of one's greatness? How much pornography to meet the need for intimacy? How much entertainment to substitute for missing play? It takes an infinite amount. That's good news for economic growth, but bad news for the planet. Fortunately, our planet isn't allowing much more of it, nor is our ravaged social fabric. We are almost through with the age of artificial scarcity, if only we can release the habits that hold us there.*

Just as people think that if only they had more money, then they would be happy, so also have we thought that we could solve the problems of the world if only we had access to vastly more energy. Oh, the things we could do!

Maybe everyone could have their own mansion and automated private jet. Then we would be happy for sure.

Tellingly, even those who do possess their own mansion and private jet are no happier than anyone else. As I wrote the earlier paragraph on depression, anxiety, etc., I looked online to make sure that I wasn't just inventing the claim that the wealthy are just as prone to these conditions as others. Isn't that how you do research too—decide what must be true and then go look for the evidence? Anyway, I came across an article in

Forbes magazine entitled "Why the Super-Successful Get Depressed."[4] Apparently, depression is rampant among chief executives. It makes sense. Before you've "made it," you can blame your unhappiness on not having a bigger house and fancier car, a yacht and a jet. Once you have those things, then what?

I need hardly argue this point; it is a truism that happiness doesn't come from more, more, more. As individuals, we get it. Collectively, we do not. Public policy takes for granted the quest for more.

There is good explanation for that: Our economic system requires endless growth. Therefore, to choose from qualitative instead of quantitative values, we will need a radically different economic system. We are not talking merely about personal choices here; we also need to change the conditions from which people choose.

This is not a call for stasis. There are many ways to develop, not all of them requiring unending increases in scale. From the standpoint of technologies of force, the currently dominant civilization is indeed the most highly developed the world has ever seen. It exerts more energy to do more work than any before it. But the civilizations it destroyed, marginalized, or absorbed were highly developed in other ways: in their understanding of body energetics, technologies of dreaming, architectural aesthetics, cultivation of consciousness, management of landscapes. Each culture explored and developed human capacities that we are hardly aware of today. Many of them seem impossible to those immersed in the Story of Separation with all its causes and categories. As that story falls apart, these capacities come back into view, and we turn instinctively toward the marginalized as a source of knowledge. This kind of development depends not at all on maintaining current levels of energy consumption. Quite the opposite: development in the Story of Interbeing can hardly happen in the present environment. Climate-controlled dwellings, digitized relationships, long-distance commutes, industrially processed food, alienation from material production, disengagement of the body from work ... all of

---

4   Walton (2015).

the things that abundant energy has made possible now impede our next phase of development.

How do *you* want to develop? Which future is closest to your aspirations: a giant flat-screen TV and a robotic housecleaning system in a five-thousand-square-foot house with a three-car garage, accessible by your own private helicopter? Or a small house of natural materials in sacred geometric proportions, ringed by gardens bursting with life, busy with birds, linked to other dwellings by footpaths in a community of people you care about deeply? You're not the only one! Do you want to "develop" your consciousness, your sensitivity to subtle energies, your familiarity with local plants and animals, your emotional intelligence, and the authenticity of your relationships? Many people hunger for this kind of development. Can you imagine what society would become if it supported it and pursued it collectively, rather than marginalizing it? Such development certainly doesn't require increased consumption of energy or materials. To the contrary—the quest for more forestalls the shift of attention and priorities that this other mode of development requires.

Transition to Abundance

If compelled to take sides in the renewable energy debate after being made to feel stupid by both, I would side with the "optimists." I think that photovoltaics in particular will grow much faster than most analysts predict, as the price of panels declines, as new storage technologies mature, and as their energy efficiency improves. Beyond that, I believe in the power of human will to bring vision into manifestation. Sustained commitment to a possibility converts it into an actuality. If humanity unites its creative powers to establish a renewable energy system, it will happen. There is currently an explosion of creativity in the field: biofuels from cyanobacteria, rail car energy storage, thermal batteries, biogas fermenters, multi-layer PV, and so on.

A renewable energy future is in reach, but let us not invest it with utopian expectations. We could transition fully to renewable energy, only to discover that it wasn't the solution to our problems after all. Switching fuels will not alter the deep preconditions for human misery and ecological devastation on this earth. So-called "green energy" can even exacerbate ecological disintegration, as the case of big hydroelectric dams and industrial biofuels illustrates. Unless we turn toward other dimensions of ecological healing—soil, water, biodiversity, etc.—the condition of the biosphere will continue to worsen. And unless we address the roots of social and psychological misery, sustainable energy will just sustain more misery.

The same admonishment applies to energy technologies that conventional opinion rejects as fanciful. Mainstream readers might be surprised to know that an entire subculture believes in so-called "free energy devices" or "over-unity devices," which draw energy from sources unrecognized by conventional science. Many in this subculture are highly educated; certainly they are not ignorant of basic scientific principles such as the Second Law of Thermodynamics. Again though, the question of the authenticity of these devices is the wrong question. Free energy devices, if they exist, are not going to save us any more than photovoltaics (another free energy device) will, or any more than petroleum did. Abundance is a state of mind and a function of social relationships. Technology is but its tool. We could have abundance right now, with no new technology, if we rid ourselves of various systems of artificial scarcity, epitomized by the artificial scarcity of money.

I have detoured into this issue, no doubt at risk to my credibility, in order to illustrate an important point: the key issue is not energy. Energy technology will not be our deliverance. Those who decry free energy devices as an escapist fantasy that distracts us from the real issue are correct even if the devices are genuine. They do not yet fit into this world. If and when they do, it will be when we have ended the War on Nature and transcended our ambition to master the world by force; we will have them, in other words, when we no longer need them to address scarcity. Their purpose is not to sustain and intensify the present mode of civilization. We are left today with more modest technologies like wind and solar, whose limits invite us to rethink the growth paradigm.

The conventional mind conceives abundance as a function of quantity, but in practice abundance depends equally on distribution and therefore on relationship. This is obvious when it comes to money, at a time of hyper-abundance for a few and poverty for the many. Since the Great Recession, the economy and money supply have grown, but nearly all of that growth has gone to the top 1 percent. Greater quantity has not meant greater abundance; the flood of central bank money has not soaked into the humus of the real economy. Similarly, total annual rainfall has increased in many places that are experiencing desertification—again, because damaged soil repels the concentrated rains of a disturbed hydrological cycle. Food too exists in theoretical superabundance on earth, yet it is so unequally distributed that nearly half goes to waste while one in five children goes hungry. In view of all this, when we consider energy abundance perhaps we should orient toward distribution and scale rather than quantity and source.

Most environmental thinkers today conceive of our civilizational transition as being from fossil fuels to renewable energy. Another kind of transition would be from a centralized system to a distributed system. Many renewable energy sources are well suited to a decentralized system. We can have rooftop solar and neighborhood biogas fermenters; we can't have rooftop coal-fired turbines or neighborhood nuclear power plants. In Africa, large regions are bypassing the construction of a power grid altogether in favor of rooftop solar.

Distributed energy is part of a larger trend of relocalization, which is necessary to bring us back into intimate relationship with the soil, water, biota, and culture of place. Like energy consumption, the trend of converting the world into standard units has nearly peaked. Whether in agriculture, economics, or technology, we need to embrace again the uniqueness of each place. That is what makes places come alive. Local, closed-loop systems must replace global mine-to-landfill systems. Certainly, some dimensions of human culture will continue to be global, just as some planetary systems are, but generally speaking, healing means the renewal of lost circles of life.

What humanity creates depends on the vision that inspires us and on the story that imbues action with meaning. To debate the viability of various alternative energy strategies focuses the conversation on too narrow a realm and draws from too limiting a story. The energy crisis, like the ecological crisis to which it is related, is occasion to move us from domination to participation. Energy then becomes a question of relationship not quantity.

Like all living beings, we will always expend energy to alter the environment, but in an age of partnership with nature, the notion that human progress depends on *more and more* energy is obsolete. How we obtain energy and how we choose to use it will both be part of a larger choice: What kind of world shall we live in?

Population

One arena where the question "Who do we want to be?" takes concrete significance is the issue of population control, where again quantitative thinking draws us into the wrong debate. I detour onto this subject mainly because whenever I post an article on some environmental topic, invariably I receive comments to the effect that I have ignored the "elephant in the room"; namely, population growth. After all, it seems obvious that no matter how green our lifestyle, the planet will be unable to support an endlessly growing population, while if we could reduce population to, say, 1900 levels, then current consumption patterns would pose little problem.

Ah, another simplifying narrative that promises salvation if only we can reduce a global number.

Like many simplifying narratives, however, the fixation on population obscures more fundamental issues. One of them is resource consumption. If everyone consumed resources at the rate of an average North American, the sustainable world population would be about 1.5 billion. If everyone lived the lifestyle of an average Guatemalan, the present population would be sustainable. And if everyone lived as ecologically as a traditional Indian villager, the planet could sustain 15 billion people or even more.[5] Most estimates put the carrying capacity of the planet at between 8 and 16 billion people, although some experts give figures diverging wildly from this range, from less than a billion to upward of 50 billion.[6]

It would sure be comfortable to stop talking about our own resource consumption, and instead limit the reproduction of people "over there." As most industrialized countries already have birth rates below the replacement rate, the population discourse puts the onus of responsibility on less developed countries.

Like GDP or CO_2 levels, what is and is not counted tends to reflect the interests of those doing the counting. Enforced through political power, the choice of what to count often tramples the interests of the powerless. Just as GDP renders other kinds of wealth invisible, just as the greenhouse gas narrative devalues natural beings that have no obvious climate relevance, so has the population scare resulted in questionable policies aimed especially at the world's most vulnerable people.

The population control movement shares close historical links with the eugenics movement. Early twentieth-century scientific consensus held that modern technology had circumvented natural selection, threatening humanity with genetic deterioration. What nature had once done for us, we humans now had to do intentionally—weed out inferior stock. The alarm was ringing and there was little time to act before it was too late.

When explicitly eugenicist ideology fell out of favor after the Holocaust, the impulse shifted onto population control policies directed at the very same people who had been targeted by eugenics. In the United States, population control was visited most vigorously upon the indigenous: sterilization rates

---

[5] I derived these figures from national data provided by the Global Footprint Network. The data for the Indian villager, I got from Demenge (2018): "Measuring Ecological Footprints of Subsistence Farmers in Ladakh," giving figures of 1.12 global hectares per capita (gha/cap) for the Alchi-Saspol region and 0.69 gha/cap for the Trans-Singe La region. Calculating from the latter figure we arrive at a possible sustainable world population of about 18 billion.

[6] See Van Den Bergh and Rietveld (2004) for a meta-analysis of human carrying capacity studies. Estimates vary widely depending on basic assumptions about technology, agricultural methods, and resource use patterns.

for Native American women exceeded 25 percent in the 1960s and 1970s, often without proper informed consent, and usually under various kinds of pressure.[7] The peak of sterilizations occurred in the early 1970s, right after the 1970 census showed Native American birth rates exceeding those of the white majority. The sterilization and birth control program was effective: by 1980 the Native American birth rate had dropped by more than half, to well under the replacement level.[8] Similar, though less comprehensive, sterilization campaigns targeted African American women, women of Puerto Rican and Mexican descent, Asians, the incarcerated, the mentally ill, and poor whites.[9]

By far the biggest population control efforts were those implemented outside the developed world, with the vigorous support of the United States government. The sordid history of mass sterilization campaigns, coerced IUD insertion, forced abortion, and other measures perpetrated almost exclusively on women of color is the subject of many books. Critiques of population control fall into two general categories: techno-utopian and post-colonial. (I will ignore the genre that fixates on the mad schemes of gibbering abortionists and fiendish U.N. conspirators.)

The techno-utopian critique rejects any limits on the ascendancy of the human race. It says that rising population is not a problem for the planet because the more people there are, the more innovation will come to bear on solving our problems. Human creativity is unlimited, it says, so any ideology that seeks to rein us in (for example, valuing and respecting all of Gaia's beings) is a kind of "anti-humanism." Robert Zubrin's article "The Population Control Holocaust"[10] offers a vigorous, if polemic, example of this genre that gives due consideration to the racist motivations and imperial geopolitical calculations behind the population control movement, which was funded and promoted by the top philanthropic foundations, think tanks, and the United States government.

---

[7]   Lawrence (2000).

[8]   Ibid.

[9]   Ko (2016).

[10] Zubrin (2012).

In advancing the premise that human ingenuity can solve any problem, Zubrin cites the "Green Revolution" that supposedly forestalled global famine and obviated the dire predictions of Malthusians like Paul Ehrlich. The Green Revolution, though, which spread mechanized, chemical-intensive agriculture across the globe, has been an ecological and social catastrophe. The gains in yield it is supposed to have produced are questionable and in any case unsustainable.[11] As described in the previous chapter, ecological farming practices can outperform industrial agriculture. In any event, hunger is ever a matter of politics and economics, not total food availability. The horrific famines in nineteenth-century India coincided with high grain exports to Great Britain. The 1974 famine in Bangladesh happened despite grain production per capita being higher than in 1973. The great Bengal famine of 1943 was caused in large part by British policies that prevented grain imports to the area and even shipped grain out of it.[12] The Ethiopian famine of 1984 came in the midst of a civil war, with food aid cutoffs and punitive crop burning in rebel areas.[13] In all these examples, drought and other natural disasters were the straw that broke the camel's back.

Global food production has long been well above the level needed to feed everyone. The quantitative mind thinks that famine must be caused by not enough food, but, at least in modern times, it has always been because of unequal distribution.[14] America is the richest country on earth, in which some 40 percent of all food is thrown away uneaten, and in which one in six

---

[11] Certainly, in a side-by-side comparison of a monocrop planting the chemically fertilized, weed-controlled, insect-controlled plot will fare much better. But when peasant agriculture as a whole is considered rather than the metric of commodity grain produced, the matter becomes murky.

[12] Horton (2010).

[13] Clay and Holcomb (1985).

[14] In earlier times bad weather and natural disasters could easily cause famines because transportation was poorly developed, so that surplus in one place could not meet deficit in another. For example, the famine of 1315 that devastated northern Europe did not affect the Mediterranean, but there existed no infrastructure sufficient to transport enough food to meet the need.

people is food-insecure. Globally the same basic truth holds: by far, enough food is wasted to feed every hungry person, even without considering the vast tracts of rich land planted in biofuels, lawn grass, and animal feed.[15] Population control is a false solution to hunger.

None of this is to say that Earth can support unlimited numbers of humans. It does illustrate that the basic problems we face are not fundamentally technical in nature. Technology in the form of the Green Revolution did not save us from famine; nor, would I argue, is technology in the form of geoengineering going to save us from climate change. Both are expressions of the cult of quantity.

Another critique of population control is rooted in deep ecology and post-colonial thinking. On one level, there is the old racial-imperial mindset of preventing the filthy heathens from reproducing unchecked. Less obvious is the imposition of modern living patterns through what Frédérique Apffel-Marglin calls "developmentalist feminism." She writes,

> *The modern bourgeois epistemology of individualism and its value of self-control transmute all those who do not live their lives in that fashion into deviant "others" who need either to be educated or, failing that, coerced into the proper, normative behavior. Professionals—therapists, educators, doctors, and so on—construct rational and individualistic models to be applied universally.[16]*

Developmentalist feminism promotes a characteristic norm of female progress: emancipation from the bondage of childbearing and village life to enter the world of professional paid employment, money being crucial to individualistic autonomy. In a modern context where community has disintegrated and women become dependents of their husbands, such autonomy is highly desirable. But in the context of the rich community life in the less developed world, emancipation simply replaces dependency

---

[15] The animal feed does end up feeding animals that are then eaten by humans, but the yield of calories and protein is very low compared to planting the land in food for humans. Meat production should not require large amounts of feed.

[16] Apffel-Marglin (2012), 147.

on community with dependency on employers and the global economic system. Women who may have had a lot of power in village society have very little in the globalized institutions to which they provide labor. The "women's empowerment" rationale for population control policies therefore buys in to a normative Western conception of what life should look like. It takes for granted that these societies should become more like our own. That is the essence of "development."

Given that the same ideology of development propels the world toward a high-consumption, resource-intensive way of life, we might want to be skeptical of this particular outgrowth of population fundamentalism.

When we reject the usual rationale for population control policies—hunger or climate change—we face another question: What kind of world do we want to live in? A strong case can be made for Earth being able to support 50 billion humans, but do we want to live on such an Earth? If we aren't forced to stop or reverse population growth, might we do it for some reason other than survival necessity?

Accustomed as our society is to war thinking, it is no wonder that the solution to population growth is birth control. Problem: population; reason: too many babies; solution: prevent babies from being born. In fact, access to contraception is a minor factor in determining fertility rates. By far the biggest influences on birth rates are (1) education of women and (2) mortality rates. The former is a proxy measure of affluence, social stability, and transition out of patriarchy. (No, it is not that dumb peasant women need to be "educated" into wanting smaller families.) As for mortality, if many children don't make it to adulthood, parents and the culture will want more of them. Rising life expectancy leads, within a generation or two, to lower birth rates.

Most developed countries today have a fertility rate lower than the replacement rate. The replacement rate is about 2.1 babies per couple (it would be exactly 2, except that some don't reach reproductive adulthood). To pick a few examples, at the present writing, the birth rate in the United States is 1.87. In Uruguay and Chile it is 1.81. In Russia, Canada, and China it is about 1.6. In Germany it is 1.44, in Japan 1.41, in Poland 1.34, in South

Korea 1.25, and in Taiwan 1.12.[17] Most of these are places with high life expectancy and low infant mortality. Meanwhile, the highest fertility rates are practically all in African countries with some of the world's lowest life expectancies. The only non-African countries in the top forty are Afghanistan, Iraq, and Palestine—places where life is uncertain.

High fertility rates will recede more quickly into history if we can end survival uncertainty caused by war and economics, along with patriarchal domination of women. Strong social ecosystems support population equilibrium, just as strong natural ecosystems support climate equilibrium. As with energy, "how much" is the wrong question. The right question is how to create the base conditions for health. In energy, we are transitioning from a high-growth, high-waste model to a steady-state model that allows other kinds of development. In population we are transitioning from a high-fertility, high-mortality growth model, again to a low-fertility, low-mortality steady-state model. Civilization is in the midst of a phase transition. Climate change and ecological limits provide the initiatory catalyst.

---

[17]   Data from The World Factbook of the Central Intelligence Agency.

10

Money and Debt

A Game of Musical Chairs

When we speak of a transition away from a high-growth, high-waste society, when we speak of nonquantitative notions of development, we immediately encounter the topic of economics. And when we speak practically of the "conditions of our choice" to continue ruining the biosphere, it is often that very quintessence of practicality—money—that determines our choice. Money is seldom the friend of earth healing. Usually, in our current system, there is more money to be made destroying an ecosystem than preserving it. Deforestation, drilling, overfishing, wetlands draining for real estate development ... the power of money drives them all. But why? Is money just bad? Are human beings just greedy? Must we forever fight a war against the money power?

The following metaphor suggests that the answer to these questions is no.

Most readers will be familiar with the game Musical Chairs. Imagine a large game of Musical Chairs is in progress, with a thousand people and 950 chairs. Everyone is dancing around the circle, and when the music stops they all rush for a chair. Those who don't find one in time are out of the game, and the next round proceeds with 950 people and perhaps 903 chairs.

Now let's make the game a little more interesting. When you lose a round, you aren't merely out of the game. You also lose your home and must choose between food and medicine for your children. The very survival of you and your loved ones is at stake. The game begins. Everyone is in a state of anxiety, maneuvering for the most advantageous position. When the music stops a mad rush ensues for the chairs, an elbowing, shoving free-for-all in which the chairs go to the strong, fast, and lucky.

Sitting outside observing the scene is an economist, a biologist, a politician, and a priest. "Will you look at that," says the economist. "That's human nature for you. Everyone out to maximize their own interest at the expense of everyone else."

"Yes," agrees the biologist. "We are watching the survival of the fittest in action. Only the strong, quick, and ruthless will survive. It's just human nature."

"Lucky thing we've got us around," says the politician, "to impose law and order to curb human nature and enforce decent behavior."

"I'm going in there to teach them to be nicer to each other," says the priest.

Is this free-for-all really human nature, though, or is it an artifact of the rules of the game? Imagine if there were one thousand people and one thousand chairs of different shapes and sizes, and the game were a matter of matching the right person to the right chair. What would "human nature" look like then? Who likes a soft seat? Who likes firm? Who has long legs? Who has a large buttocks? The play of the game would look very different; it would involve a lot of communication and cooperation. Different structures would emerge to match the right person with each chair. There might still be some competition, but it wouldn't be baked into the rules of the game.

Structures might emerge in the original game too. Sometimes a strong person might secure a chair for himself and a friend or two. Small groups might form to secure chairs at the expense of other groups. Certain altruistic individuals might sacrifice their own chance at a chair so that the young mother with the baby can get one. Others (after having secured a chair themselves) might exhort others to be a little nicer to each other and not push so much. The rules of the game, though, entail that generosity equals self-sacrifice. More for you is less for me. It is a zero-sum game; in fact it is a negative-sum game.

Musical Chairs is closely congruent to our current economic system (with one important difference I will discuss shortly). Because money in our system is lent into existence, and because those loans carry interest, at any given moment there is always more debt than there is money. Just as in Musical Chairs, everyone is therefore set into competition with each other for never enough money. The "strong, quick, and lucky" get a "chair"—the money they need to enjoy material security—and the weak, the unfortunate, and the disadvantaged do not.

Imagine further that the circle of 950 chairs is not evenly arranged. Some areas have a sparser distribution of chairs than others, and it is into those sections that the black, red, and brown people are put. They long for a place at the denser part of the circle, and perceive racism to be the source of their poverty. What they do not realize, though, is that if not them, *someone* has to be deprived of a chair, because that result is designed into the rules of the game. To the impoverished black, red, and brown people it sure looks like racism is the cause of their poverty, but in fact it is more of a symptom and enabler of a system that must impoverish someone: if not us, them. So different factions seek to engineer the circle so that their section has proportionately more chairs (in the real world, this corresponds to the imperial control of resources), engendering a new level of competition and setting conditions conducive to racism, nationalism, and imperialism.

Everyone is so focused on winning a higher proportion of chairs for themselves and their group that they don't question the rules of the game and whether those might be changed.

Here's a little snapshot of how it plays out. According to Matthew Desmond's piercing book *Evicted,* tens of millions of people in America today spend 50 percent or even 70 percent or 80 percent of their income on rent, living always just one health crisis or car repair away from a downward spiral that starts with eviction and ends in total destitution: family breakup, prison, homelessness, or worse. One would like to blame heartless landlords, but in fact the heartlessness originates systemically. It is a system predicated from top to bottom on the maximization of self-interest, and therefore demanding we treat others as instruments of our own utility. The "heartless landlord" is subject to the same basic economic insecurity as everyone else, albeit more distant. An economic downturn, a stock market crash, and she too might descend into penury. Quite often, the building owner is a real estate holding company that contracts out building management and is under pressure to maintain a decent rate of return in order to service debts to its own creditors. At the base of the system are the large institutional investors, depriving any firm of capital unless it produces the highest possible return. Maybe we should blame *their* greed—except that the largest are mostly pension funds, desperate for high enough returns to fund the retirement of teachers, firefighters, and other workers.

Heartlessness inevitably accompanies this system of dehumanization and exploitation. If you care too much, quite often you'll go out of business. No chair for you.

Leftists love to blame corporations and their executives for the planet's predicament, but they are creatures of the system, carrying out structurally determined roles. Certainly there is some leeway for corporations to act more or less in the public interest, but to deviate too far from aggressive service to the bottom line faces the company with the iron law of the marketplace. Its more ruthless competitors will destroy it, or perhaps squeeze it into a small niche. That is why it is foolish to hope that ethics training or meditation in the boardroom will fundamentally change aggregate corporate behavior.

The movement to reform corporate charters (to mandate limited terms and profits, and require they serve the public benefit) is a step in the right direction, but we also must understand that the corporation as we know it is not some unfortunate accident of history, but a natural adaptation to the "rules of the game" as well as the inevitable end product of an all-encompassing story. It is like the fulfillment of a prophecy. The doctrine of rational self-interest, which is not true, has never been true, and never will be true for human beings, was the prophecy. The corporation is the vehicle of its fulfillment. It is the culminating expression of the ideology of self-interest.

The behavior of corporations is a distilled and expanded version of how people generally behave in an economic system of artificial scarcity that distances causes from effects. The corporation is a more ruthless player of Musical Chairs than most actual humans can be, and therefore excels at the game. Nonetheless, you and I replicate the same basic ruthlessness whenever we seek the best deal. Suppose you want to fill your gas tank. One gas station charges ten cents per gallon more than the other, for exactly the same quality and service. Everything you can measure is identical. Which do you choose? Do you think, "Well, if I choose the cheaper one, it won't make a fair profit or be able to compensate its employees fairly. I'll choose the more expensive one." Probably not. Writ large, this is essentially the behavior of corporations. They are the highly developed historical implement of the subordination of life and matter to the pursuit of self-interested value. Like so much of what we blame for our woes, they are more a symptom than a cause of the present crisis.

None of this is meant to dissuade companies and individuals from doing their best to adopt sustainable practices. Such choices contribute to a shift in the story of "normal." Even when they appear to fail, they bring into sharper relief the conflict between our system and our ideals. Although better ethics or more spirituality won't change the rules of the game or reverse the tide against which sustainability must swim, I actually see value in the oft-derided "corporate mindfulness" movement, "conscious capitalism," and meditation in the boardroom. Critics say they merely facilitate business-as-usual under a cloak of spirituality, which would be true indeed if these practices had no real power. Applied sincerely, they have the opposite effect: they make business-as-usual harder. They arouse troublesome questions with no easy answers. They precipitate a crisis for the organization and the people in it. And that is good.

The Growth Imperative

There is one key dissimilarity between Musical Chairs and the money system, although this difference does not alter the fundamental pressures that generate anxiety and competition. The dissimilarity is called "economic growth." For illustration, imagine a simplified economy in which the banking system has lent a million dollars at 7 percent interest out to a thousand people. Each person has received $1,000 and after ten years' time must pay back $2,000. It is mathematically impossible for more than half of them to do it, because only $1 million has been created in the first place, but $2 million is owed. If that were all to the story, a ferocious competition would ensue in which at least half of the people would go bankrupt, just like the Musical Chairs game.

In fact, when the initial loans of $1,000 each come due, nearly everyone can pay them back. Why? Because in the interim, the bank has lent that additional $1,000 per person into existence—another $1 million in total debt. (The money isn't lent out evenly, though. It goes only to those who the bank thinks will pay it back with interest.) As long as new money is continually created (lent into existence), the system can keep running. When lending slows down—or even fails to grow at the pace of the rate of interest—then bankruptcies are inevitable. A vicious circle looms: layoffs and pay cuts, leading to falling demand and falling prices, leading to falling profits, leading to fewer lending opportunities, leading to more bankruptcies and layoffs. This is called a depression.

To forestall depression, economic growth must never end. It is no accident that, because they take the current system as unchangeable, politicians left, right, and center all champion economic growth. They disagree about how to accomplish it, but everyone agrees that it is necessary. They are right—under the present financial system, it is necessary.

Economic growth is rarely questioned by any politician, since its premises are so deeply entrenched. That is why the progressive left conjures fantasies of "sustainable growth," imagining that somehow we can continue to replace relationships with services and nature with products without ecological or social depletion.

Numerous economists are arguing—not only for ecological reasons—that we are nearing the limits of economic growth.[1] Unfortunately, our present money system works only in the context of rapid growth, essentially because positive returns on capital investment are necessary systemwide in order to motivate sufficient lending. Lending, in turn, is the basis of money creation. Without new money constantly being created, the means to service the debts is diminished, resulting in bankruptcies, unemployment, concentration of wealth, and the need for austerity to temporarily service the debts when rising income cannot. This generates relentless pressure on governments to find new ways to facilitate economic growth: colonialism, exploitation of natural resources, and so on. Today we face limits to growth, leaving austerity as the only option to keep the debtors paying a while longer.

Economic growth means the growth in goods and services *exchanged for money.* Therefore, a remote village in India or a traditional tribal area in Brazil presents a big growth opportunity, because the people there barely pay for anything. They grow or forage their own food. They build their own houses. They use traditional healing methods to treat their sick. They make their own music and drama. Imagine the development expert goes there and says, "What a tremendous market opportunity! These backward people grow their own food—they could buy it instead. They cook their own food too—restaurants and supermarket delis could do it for them much more efficiently. The air is full of song—they could buy entertainment instead. The children play with each other for free—they could enroll in day care. They accompany adults learning traditional skills—this society could pay for schooling. When a house burns down, the community gets together to rebuild it—if we can unravel those ties of mutual aid, there's a

---

[1] For a case that the age of economic growth is over that does not rely on ecological arguments, see Gordon (2012).

big market for insurance. Everyone has a strong sense of social identity, a strong sense of belonging—they could buy brand name products instead. Everyone is joyful and content—they could be buying a semblance of that through legal and illegal drugs and other forms of consumption."

Okay, I'm getting a bit dizzy with visions of riches, but you get the idea. The question is: how are these people going to pay for all that? Easy. They earn money by converting local natural resources and their own labor into commodities. The rainforest becomes a palm oil plantation. The mountain becomes a strip mine. The river becomes a hydroelectric plant. The population abandons their traditional ways and goes to work in the money economy. A few become doctors, lawyers, and engineers. The rest migrate to the slums.

In a nutshell, this is the process called "development." It is what development loans have funded for more than half a century. It accompanies an ideology that says that money equates with well-being, that development along the model of the West is a good thing (or an inevitable thing), that a high-tech life is superior to a life close to nature. These assumptions are difficult to refute using logical arguments. Usually, shedding them requires spending time in less developed cultures, witnessing the joy and depth of aliveness there, and seeing their beauty erode as they modernize.

The word "development" contains a value judgment that others are less advanced along an implicit scale of progress. Accordingly, a financial system that compels development is a good thing. And it looks like a good thing indeed if one accepts GDP as a valid measure of well-being. When tens of millions of Indian farmers switched from biodiverse organic agriculture for local consumption to the chemical-intensive, water-intensive monoculture production of crops for commodities markets, their contribution to measured GDP rose markedly. Why? Before commoditization, much of the food was eaten by the extended family that grew it, and circulated through the community through nonmonetary systems of reciprocity. The rest was sold in local markets in an informal economy. Transitioning to chemicalized, mechanized agriculture involved going into debt to buy machinery, fertilizer, herbicides, insecticides, and seeds. The resulting suffering that happened when commodity prices fell and farmers couldn't make debt payments is well known: banks foreclosed on land that had been in families for centuries, even millennia, and hundreds of thousands of farmers committed suicide. The younger generation had no choice but to move to the burgeoning megacities as traditional livelihoods evaporated. Industrial products replaced the output of potters, toolmakers, weavers, and other artisans. And GDP figures rose.

Usually, blame for this state of affairs is placed on Monsanto and the banks, invoking the bugbear of corporate greed. It is sure nice to have something to hate and to blame, and it is true that Monsanto aggressively pushed its chemicals and GMO seeds in India. We must understand though that this corporation swims in the ideological waters of "modernization" and believes itself to be rendering a great service to humanity. Yields have gone up and the economy is growing. We are helping backward peasants enter the modern age and feed the world's hungry masses. As usual, the problem with these justifying numbers lies in what isn't measured, such as:

- Social disruption caused by abrupt changes in local economic patterns

- Lost subsistence food crops that never entered harvest tallies

- Dietary diversity, which is important to good health

- Future losses due to falling water tables due to water-hungry monocultures

- Future losses due to soil erosion caused by modern agricultural processes

- Contamination of soil and water by chemicals

- Losses stemming from long-term soil compaction and loss of mycelia

- Future effects of superweeds and insecticide-resistant insects

As is so often the case, what is most important is what the numbers leave out. The story of modernization that Monsanto inhabits depends on the invisibility of these things. That same invisibility is what keeps natural human compassion from operating, as suffering hides behind the numbers. It isn't just Monsanto of course; the mindset of development is intrinsic to the system we live in. It is part of the Story of Separation with its thread of humanity ascending to lordship over nature. We might even admire Monsanto for being an especially innovative practitioner of a near-universal ideology. To blame Monsanto's greed (or that of its brethren like Syngenta, DuPont, Dow, Bayer, etc.) is to misdiagnose the problem, or at best, to attack the symptom rather than the conditions of the disease. The conditions are the story and the system. From within that, those working for Monsanto see themselves as the good guys and the anti-Monsanto marchers as delusional hippies who just don't get it. They just don't understand that thousands of dedicated scientists—scientists!—have devoted their careers to advances in crop science that are bringing such benefits to the world. They don't understand that we are in a race against hunger.

Understanding the system and the story that Monsanto etc. operates in, we can direct our activism toward changing the system and rewriting the story. Even on those occasions where a fight is necessary, we will be much more effective acting from a full realization of how the opponent sees the world and itself.

System and story are deeply entwined. In case the ideology of modernization and development isn't enough, there is intense financial pressure to conform to it. In the banking analogy above, the new $1,000 isn't lent randomly to just anyone. It is lent to those who are deemed likely to be able to pay it back with interest—by earning money from someone else in the circle. Money originates as credit to those who will pay it back with interest, ultimately from participating in the creation of new goods and services. That means the conversion of social relationships into services, and natural wealth into products. That is called development.

Development and Debt

The ideology of development has a justification in classical economics that says that the outsourcing of various life functions to specialists, aided

230

by technology and mass production, allows human needs to be met more efficiently. It is true that modern society produces vastly greater quantities of wealth than traditional ones—when that wealth is measured in money. Economics assumes that more money equals more happiness; that the more goods we can buy, the more "good" life is. This logic is valid only to the extent that human needs and desires can be met by things that can be quantified, bought, and sold. We in the dominant culture have more of those things than ever before; meanwhile, much of what deeply nourishes the human being is artificially scarce. We have a scarcity of time, scarcity of beauty, scarcity of intimacy, scarcity of real connection to community and to nature. Thus deprived, we are always hungry for something, but no amount of money, possessions, status, cars, social media likes, or domestic floor area can meet these unfulfilled needs. Those aren't really what we are hungry for. We call this endless hunger "greed" and rail against it as the cause of our present social and ecological nightmare, but as usual, we are warring against the symptom not the cause. Greed is a symptom of scarcity. What we call development has cut us off from real wealth. It has distanced us from place, from people, from other-than-human beings, replacing those relationships with standardized, mediated relationships. In the course of development, we have gained much of what we measure. But what have we lost?

As the Story of Interbeing suggests, a whole human being is one held in a close web of intimate relationships. To cut off these relationships is to amputate part of the self. To restore wholeness and assuage the hunger we call greed, we must therefore restore those lost relationships. This means rebuilding community, reconnecting with our food sources, and in general interacting with nature as a participant rather than as a spectator. It means reversing certain key aspects of development. It does not mean abandoning technology or global culture, but finding their proper place, reclaiming the territory they have usurped, and embracing a different conception of progress. In this conception, which emphasizes the qualitative dimensions of life and recognizes forms of wealth that our society ignores, we no longer see ourselves as more "developed" than agrarian villagers or Amazonian hunter-gatherers.

A friend runs a farm and retreat center in Brazil. He needed more accommodations for visitors, so he hired some nearby indigenous people to build a new building. "I didn't hire them because they are Indians," he said. "I hired them as architects."

Using no measuring equipment, no metal fasteners, and no materials except what they procured from the land, in just three weeks these indigenous builders constructed a dwelling that could accommodate forty people in hammocks. It is a marvel of intelligent design, cool in hot weather, warm in cold weather. Smoke from the central fire pit rises quickly to be released through the permeable roof, proofing it against insects; yet it is fully waterproof. Its utilitarian genius is matched by its aesthetic perfection: despite the lack of measuring equipment, its dimensions are in precise golden mean ratios; moreover, the building conveys a striking presence and aliveness (judging from the photograph). My friend says that when professional architects visit the structure they sometimes weep in humiliation, so far beyond their capabilities it is.

In a truly advanced society, everyone would live in a building that beautiful.

A new kind of progress in a degrowth economy does not mean a regress in well-being. It might mean a regress in certain quantitative measures of wealth that we accept today: fewer units of floor space, automobiles, and energy consumption per capita. It might mean more herbal medicine and less pharmaceutical medicine, more bodywork and fewer high-tech medical procedures, more beautiful buildings and fewer big ones, more singing and fewer music purchases, more time outdoors and less in gyms, more free play for children and less time spent in organized activities. Childhood was not always so expensive.

The deemphasis of the measurable corresponds to the transition away from the conversion of nature into commodities. It allows us to see the beings of nature, the ecosystems and species, as sacred beings in their own right. Beauty and sacredness and love tend to get lost in the numbers, which put a finite value on the infinite. In both economics and ecology, we need to shift to values that cannot easily be measured.

Development has happened over centuries in the West, where there is hardly anything we do in community anymore, and hardly anything that has not become a product or a service. Because there is so little left to convert into money, the system reaches to "less developed" parts of the world to maintain overall growth.

Development loans finance the infrastructure to extract natural resources and haul them away, and to build industries to make local labor available to global corporations. Moreover, the pressure to repay the loans (with export-derived dollars or euros) ensures that that infrastructure will be used for its intended purpose. Since loans to less developed countries bear a higher rate of interest, overall return on investment remains high enough to keep the financial system functioning. In essence, growth is imported into the developed countries from the less developed.

Furthermore, because their debt is at hopelessly high levels, the pressure to "develop" never ends. If ever the pace of resource extraction or labor market opening falters, then the creation of new financial wealth lags behind loan payments, and the country must cannibalize its existing wealth instead to pay creditors. That process is called "austerity." Creditors demand that the debtor nation privatize public assets, slash pensions and salaries, liquidate natural resources, and cut public services so it can use the proceeds to pay creditors and avoid default. Another term for avoiding default is "staying in debt forever," since these debts are unpayable. According to a report by the Jubilee Debt Campaign, since 1970 Jamaica has borrowed $18.5 billion and paid back $19.8 billion, yet still owes $7.8 billion. The Philippines has borrowed $110 billion, paid back $125 billion, and owes $45 billion.[2] These and many other countries might actually have a positive balance of payments if it were not for their interest burden, which essentially condemns them to render tribute unto the global financial system forever, exporting more than they import for the privilege of staying in debt.

---

[2]   Dear et al. (2013).

The same pressure afflicts developed countries and individuals as well. In an environment of artificial scarcity, we are all under pressure to orient our lives toward the production of salable goods and services, just as less developed nations are. Those nations must constantly scramble to find some way to keep making payments on their debt. Does that sound familiar? It is the same for individuals. If you can't make payments from rising income, then you'll have to implement a personal version of austerity: selling off assets, cutting back on health and leisure, gearing your whole life to the production of money.

Hypocrisy

What do you think is the main impediment that keeps you from devoting your life to ecological or social healing? For many people, it is the necessity to make money in the very economy that is consuming the world's social and ecological capital. Ultimately, that is where the money *comes from:* it comes from investment credit extended to whoever will create new goods and services. Those firms hire employees who will help them achieve that goal. If your goal is at odds with that—if your goal is not to help turn nature into products and community into services—then you are going to have a hard time finding a job, because generally speaking, that's where the money comes from.

There is no law of physics that says we must create money the way we do. Money is an expression of a social agreement about what is valuable and where to apply our collective will. We could choose to create money in a way that directly values and supports ecological and social healing. To some extent we already do this—central banks, which are quasi-public institutions, purchase government debt that funds programs. However, as interest-bearing debt it still feeds the growth imperative. An alternative would be zero-interest, negative-interest, or government-issued "positive money" issued in direct subsidy of social or environmental healing.

Environmentalists will always be fighting an uphill battle unless and until we change our financial system. Railing against corporate greed misses the point. Corporations are only acting out a systemic imperative. Railing against consumer hypocrisy misses the point as well. Can you blame the pollution in a coal town on its residents' personal lifestyle choices? To the contrary, if there is any hypocrisy it is in maintaining a system that compels ecocidal behavior, and then blaming people for carrying out ecocidal behavior.

In 2007, the nation of Ecuador asked the world to help it conserve the Yasuni rainforest, which some say is the biologically richest in the world. Unfortunately it is also rich in oil deposits—more than $7 billion worth lies beneath it. So Ecuador's President Rafael Correa said his country would forgo half that amount if international donors would pay Ecuador the other half through a U.N.-administered fund. The funding didn't come through—less than 1 percent was ever pledged—despite the strenuous efforts of the Ecuadorian government to lobby for the idea. In 2013 they gave up and announced plans to develop the region. Following an intense but fruitless campaign by the indigenous inhabitants to stop it, in 2016 a consortium of Chinese oil companies was given development rights. Oil drilling commenced in 2017.

Who is to blame here? Is it Ecuador for failing to protect its rainforest? Or is it a hypocritical system that on the one hand says "Preserve your rainforest to conserve biodiversity and stop climate change" and on the other hand says "But we'll give you money if and only if you chop down the forest to make way for oil rigs"?

Generally speaking, we are all in the same boat as Ecuador. We are made to feel guilty for our world-destroying activities, even as we are lodged in a system that virtually compels us to participate in world-destroying activities.

Failing to understand the systemic nature of hypocrisy, we might condemn the climate activist who calls for an investment boycott on fossil fuel companies one moment, and fills her tank with gas the next. I might condemn myself for lamenting the procurement of conflict minerals from central Africa, using a computer that likely contains those very minerals. This is not about pursuing personal purity. It is about understanding causes.

Like greed, hypocrisy is a false enemy. It is another symptom. I highlight the symptom only to illuminate the disease.

Hypocrisy is a symptom of a double bind. When you notice hypocrisy in yourself or others, instead of swelling with indignation or judgment, you could take it as a sign that the hypocrite is in an impossible situation. In the classic Batesonian sense, a double bind presents a person with conflicting imperatives, each operating at a different level of abstraction or awareness. An ecocidal economic system puts us in a double bind: success in meeting one imperative (personal security) means failure in meeting another (serving planet Earth). The resulting discomfort encourages the pretenses and self-delusion that constitute hypocrisy. That is how we often cope with a double bind, but there is no escape other than to overthrow its premises.

It might be helpful to illuminate hypocrisy as a means to reveal the double bind beneath it, to say, in a spirit of alliance, "Let us do something about this dilemma together." But attacking people or corporations for their hypocrisy does no good whatsoever. If the underlying dilemma is not addressed, at best they will make superficial changes to avoid the *appearance* of hypocrisy.

So both greed and hypocrisy—the favorite targets of environmental righteousness—share a common root in our economic system. If greed drives the consumption of nature and if hypocrisy enables it to continue, then we must transition to a different kind of economic system, one that no longer generates greed by separating us from the real wealth of human and natural relationship, and that no longer feeds hypocrisy by facing us with an impossible dilemma.

The macroeconomic equivalent of this hypocrisy resides in aforementioned notions of "green growth" and "sustainable development." Here is another double bind: our system depends on growth to function, yet infinite growth is impossible on a finite planet. We must overthrow the premises of this dilemma and change the monetary system that compels growth.

Elements of an Ecological Economy

My 2011 book, *Sacred Economics,* attempted to describe what a steady-state or degrowth financial system might look like, and how we could realistically transition into it. Its main pillars were: negative-interest money creation, universal basic income, internalization of ecological costs, economic relocalization, and, animating all of these, the recovery of the spirit of the gift as the basis of human economy, creativity, and livelihood. In retrospect I have doubts about the title of the book, which, although true to its content, put it outside the notice of most practicing economists and policymakers. Happily, many of the ideas I wrote about are ideas whose time has come. The age of growth is ending despite our every attempt to maintain it; social and ecological limits press in on the economic system and it creaks under the strain. As its demise comes into view, as the insolubility of the crisis becomes obvious, ideas that once seemed fanciful are filtering into the mainstream. I will mention a few in brief that I explored in depth in *Sacred Economics* that have particular relevance to ecological healing.

Debt Cancellation. As I have described, the global debt regime is a core driver of environmental destruction. Although it seems like an unalterable aspect of reality, debt (like money itself) is a social construct, only as real as the laws and agreements that enforce it. Laws can be changed. Agreements can be repudiated. In the end, debt depends on political power.

In principle, the central banks of the world could simply purchase and annul all student loan debt, consumer debt, mortgage debt, and sovereign debt, because central banks (like the Federal Reserve) have the power to create essentially unlimited amounts of money. They could also partially cancel those debts or reduce the interest rate to zero. They lack the political mandate to do so, but we must recognize that the current debt regime is not an unchangeable part of physical reality. It is in our power to change. We do not need to be stuck in a usurious world. In the last ten years we have seen a series of bailouts that were actually "creditors' bailouts," designed to keep the debtors owing. We could change course and meet the next crisis—coming soon!—with a debtors' bailout instead.

A debt resistance movement is growing worldwide that recognizes the unjust origins and onerous effects of much debt today. A debt strike (refusal to make payments on debt) by an organized minority of debtors (individuals and nations) would quickly bring the system to its knees, since it is already so highly leveraged. Given the role of debt in driving the world-destroying machine, movements like the debt jubilee are also forms of ecological activism.[3]

Negative-Interest Money. An interest-based system is profoundly unecological. It attaches a world that is cyclic to a token of value that grows exponentially. It values the present over the future by encouraging the discounting of future cash flows. It requires endless growth in a world that is finite. These are among the reasons to investigate a system that reverses the effects of interest.

The way to apply this idea is through a liquidity fee on bank reserves. Basically that means if a bank keeps excess reserves and doesn't lend them, those reserves shrink by a rate of perhaps 5 percent a year.[4]

In that context, banks would have an incentive to lend at zero interest or even less. Lending would no longer depend on aggregate economic growth. A break-even business that would never have to expand its revenues to meet interest payments would be a viable investment. The realm of paid goods and services would not need to constantly expand. The money system would no longer drive the conversion of nature into property and product. Conservation would no longer be wading upstream against the current of financial logic.

A negative-interest money system

- Allows money to circulate in the absence of growth

- Reverses the concentration of wealth that the present system encourages

- Shifts taxation away from income and sales, toward money itself (and other rent-generating property such as land)

---

[3] For more on the philosophy, economics, and politics of debt resistance, read my article "Don't Owe. Won't Pay." in *YES!* magazine (Eisenstein, 2015b).

[4] Negative interest can be implemented even more easily in digital currencies, and could mitigate the hoarding and speculation that plague them.

- Offers debt relief without bringing down the whole system and ruining small savers

- Aligns money with the spiritual principle of impermanence and the ecological law of return

- Reverses the discounting of future cash flows, discouraging the liquidation of irreplaceable natural capital

The reader will have many questions. How will banks make money? Would it cause inflation? What is to stop speculative bubbles as people flee cash for commodities? Wouldn't it encourage overconsumption? Most of these are addressed in chapter 12 of *Sacred Economics,* which describes the history, theory, and application of negative-interest currency in detail.

I mention the idea here to hint at an alternative to the usual polemics around "capitalism," which is the usual culprit in a Left analysis of the environmental crisis. But as in so many debates including climate, the polemics obscure deeper assumptions shared by both sides. The nature of capitalism depends on the nature of capital. And the nature of capital—in particular, who owns it and what they can do with it—depends on social agreements that are not black and white, but admit many shades and variations. Negative interest turns capitalism upside down.

Socialism is usually defined as "public ownership of the means of production," but what is ownership? It is never an absolute subjugation of object to subject, as the Story of Separation would have us think. It always rests on a social arrangement. The objects themselves, the land, the water, the minerals, the trees, do not themselves admit to being owned. Even the most stalwart Libertarian doesn't think that owning something confers the right to use it to harm others. The rights of property are always socially circumscribed. The question then is: as our understanding of what harms the other evolves, what is the appropriate social agreement about who can use what for what purpose?

Activists, we will never get very far by pinning hope on inciting the public to tear down capitalism. Nor will we get very far by leaving the current system of capitalism intact. We must alter its foundations, the basic perceptions and agreements that define the categories of money and property. Even the word "mine" seems obsolete, as we begin to understand ourselves as interbeings, and as we begin to see the objects of ownership as subjects themselves. Today, we know it is wrong to own a human being—slavery is only thinkable if we dehumanize the slave. Now it is starting to seem similarly wrong to own land. We can be its steward, its caretaker, its partner, its ally, even its servant ... but its owner? How dare we?

The challenge is how to translate that understanding into our economic system. Many of us want to live more humbly and respectfully. We don't want to profit off the suffering of the other. This rising consciousness is misaligned with the current system of money and property. Negative-interest money is a step toward realigning economy and ecology.

Internalization of Ecological Costs. Today it is quite obvious that money is usually the enemy of sustainability. There is a lot of money to be made by extracting resources, clearing forests, depleting oceans, and emitting pollution. At present, there is little to be made by greening deserts, restoring wetlands, protecting habitats, or avoiding pollution. That means that government policies—and our own good intentions—must fight the money power in order to maintain a livable world that honors life.

Is this a necessary state of affairs? Does it reflect an eternal battle between altruism and selfishness, between spirit and matter, good and evil, God and Mammon? Some economists think it need not be this way, if only it were possible to make environmentally destructive activities very expensive and restorative activities remunerative. The idea is that pollution, deforestation, and so forth are a form of stealing from society, nature, and future generations. No one should be allowed to profit by externalizing costs onto someone else. Green taxes and cap-and-trade schemes for pollution rights seek to internalize these costs and align the best business decision with the best ecological decision. On the restorative side, the concept of "valuing ecosystem services" seeks to pay people to preserve land, plant forests, protect watersheds, and so on.

I have already criticized this idea on both theoretical and practical grounds, especially when it comes to reducing ecological health to a single monetized metric of carbon equivalents. We can measure only what we can see, so anything outside our cultural blinders will escape our accounting. Moreover, we unwittingly import our invisible biases into our choice of what to measure and how to measure, biases that tend to be aligned with the financial interest of the institutions and systems that apply them. What is visible to us in the civilized world? Tons of CO_2. Hectares of forest cover. Concentrations of ground-level ozone. Acidity of oceans. Numbers of species. In the service of these *measurable* things, we are willing to sacrifice what is invisible or unimportant to our eyes: generations-old social practices that allow traditional people to coexist with the land; the integrity of sacred sites; complex ecological dependencies that we have not yet learned to see or measure.

On the other hand, clearly we cannot persist in upholding a system in which profit and ecology are opposed. Can we fix the concept of ecosystem services? In fact, some of the programs justified by the concept of ecosystem services have been successful, and we should not dismiss these successes on dogmatic grounds. Farmers in Bolivia are paid to protect their watersheds, and loggers are paid to cease clear-cutting. Cap-and-trade systems for sulfur dioxide have curtailed acid rain. Learning from the failures (such as the dismal results of carbon credit trading) and carrying forth the successes, we might develop more and better ways to align money with ecology. For example:

- We can eliminate hidden subsidies that make local and ecological practices uneconomic. This is certainly the most important measure, because so many unsustainable practices are viable only because of public subsidies. Long-distance trucking companies, for example, don't pay to build and maintain the highways. This cost is borne by the public. Nor do oil companies bear the costs of imperial oil wars.

- We can use quota systems, green taxes, or auctions to limit renewable resource use to the amount that can be sustainably replenished.

- We can do the same to limit waste emissions to a rate that the rest of nature can process.

- We can also pay countries like the Democratic Republic of the Congo, Ecuador, and Brazil to preserve their rainforests, setting that amount at a level sufficient to offset the profits that would otherwise accrue from liquidating those resources.

- We can pay farmers to practice regenerative agriculture.

- We can cancel Third World debt in recognition that much of it was incurred for the purpose of extracting resources whose environmental costs were never compensated.

The organizing principle here is not to totalize economic logic. It is that people and nations should be able to make as much money from the alternatives to extraction (past a sustainable level) as they do from extraction itself. It would be hypocritical otherwise: to say, "Don't cut down those trees—but I'll only pay you if you do." Money, after all, is an expression of what society values. As what we value shifts toward ecological healing, we need to change the economic system to reflect that.

We should not pretend, though, that the financial incentives we assign toward environmentally desirable outcomes can truly represent the value of the land, water, biodiversity, etc. It is surely a good thing to align money with ecology, but we must do that without reducing ecology to money, nature to commodity, the infinite to the finite, the sacred to the profane, quality to quantity, and the world into a pile of instrumental stuff. Detaching financial incentives from the doctrine of value frees us to apply them flexibly on a case-by-case basis that fully recognizes their social context.

Universal Basic Income. At first glance, the notion of a guaranteed income for all might seem to promote consumption rather than sustainability. Actually, it could liberate people from compulsory participation in the extractive economy and free them up to serve as healers, artists, peace workers, and ecological stewards.

Universal basic income draws intense criticism from both the right and the left. The right says that without being forced to work for a living, most people would cease to contribute to society. Who would drive the buses, wash the dishes, and clean the toilets? The Marxist left says that UBI preserves the basic structure of capitalism (private ownership of the means of production); at best it would blunt the edge of capitalism's worst excesses.

While a thorough exposition of the arguments for and against UBI is far beyond the scope of this book, responding to the above-mentioned criticisms will serve to illuminate UBI's potential. First is the matter of "no incentive to work." That draws from a particular philosophy of human nature; namely, that humans are driven by rational self-interest, and won't choose to contribute to something greater than themselves unless bribed or compelled to do so. That means that you, my dear reader, would happily retire to a life of tennis and golf, *World of Warcraft* and HBO, partying and debauchery, if only you could get away with it. Fortunately, you are compelled to make a living instead.

What I see in the world though is the opposite. It is that people have a compelling desire to contribute meaningfully to the well-being of society and the planet, but that the pressure to earn a living prevents them from doing so. Or they must struggle against economic pressure to do what the world most needs right now. This suggests a malfunction in our economic system, which, ideally, is supposed to encourage precisely those things that serve the world. Instead, it encourages those things that serve the program of growth, domination, and conquest, the Ascent of Humanity. These goals no longer confer meaning and fulfillment to most of the people who serve them.

In a sense then the criticism from the right is accurate. Society as we know it would fall apart if we could no longer bribe and coerce people into performing degrading work. On the soul level, it can be just as degrading to work in high finance as to drive a shuttle bus—maybe even more degrading. In the context of UBI, corporations and entrepreneurs would have a strong incentive to design fulfilling jobs, because they would no longer be able to count on desperate people willing to do practically anything.

As for the Left critique, one might look at UBI as in fact a restoration of the commons to public ownership. The term I used in *Sacred Economics* for UBI was a "social dividend"—each person's share in the collective wealth of the world, natural and cultural, to which no person has more of an intrinsic right than anyone else. This view is especially compelling if the UBI is funded through levies on accumulated wealth (e.g., negative interest, Georgist land taxes, etc.), which essentially negate the profits available through mere ownership of resources.

Any debate about capitalism depends on the nature of capital. Both money and property exist by convention. They are stories, systems of meaning and agreement. Stories can be changed. The current Story of Money is part and parcel of the Story of Ascent; it is the foundation of a social system that is devouring the world, a system that converts quality to quantity, nature to commodity, soil into dirt, trees into board feet, and values into value. It is a system that chews up beauty and spits out money. It is the tide against which every environmental cause swims. To change it is no small matter. Money as we know it infiltrates our understanding of who we are and what is real. It cannot change unless everything changes, and when it does change, everything else will change with it. Those who see the climate crisis as a portent of a wholesale transformation of our civilization should understand, then, the necessity of that change reaching to the level of money.

11

An Affair of the Heart

In Science We Trust

If money is the keystone of the arch of modern society, the foundation is surely science. When someone demands we be realistic, often they are referring either to money, or to scientifically verifiable fact. Science provides our culture's main map of reality. If climate change indeed faces us with an initiation into a new phase of human civilization, then we might expect that science, like money, will undergo a profound metamorphosis.

Except among the religious fringe, science is a primary locus of authority in our society: for at least a century to be "scientific" has been among the highest sources of legitimacy in business, government, medicine, and many other fields. Even those who consciously reject some of science's teachings aspire to it. As our culture sees science as its foremost means to discover truth, to reject what science says seems the epitome of irrationality, tantamount to a willful denial of truth itself.

I have argued that the portrayal of the climate debate as a clash between the forces of truth and the forces of deception leaves a lot out. This is no mere battle between the intelligent and the stupid, the backward and the advanced, the corrupt and the ethical. The rejection of science, or at least of "what science says," portends a tectonic shift in the bedrock of civilization as we know it.

Science in our culture is more than a system of knowledge production or a method of inquiry. So deeply embedded it is in our understanding of what is real and how the world works, that we might call it the religion of our civilization. It isn't a revolt against truth we are seeing; it is a crisis in our civilization's primary religion.

The reader might protest, "Science is not a religion. It is the opposite of a religion, because it doesn't ask us to take anything on faith. The Scientific Method provides a way to sift fact from falsehood, truth from superstition."

In fact, the Scientific Method, like most religious formulae for the attainment of truth, rests on *a priori* metaphysical assumptions that we must indeed accept on faith. First among them is objectivity, which assumes among other things that the formulation and testing of hypotheses don't alter the reality in which the experiments take place.[1] This is a huge assumption that is by no means accepted as obvious by other systems of thought. Other metaphysical assumptions include:

- That anything real can in principle be measured and quantified

- That everything that happens does so because it is *caused* to happen (in the sense of Aristotelian efficient cause)

- That the basic building blocks of matter are generic—for instance, that any two electrons are identical

- That nature can be described by invariant mathematical laws

Philosophers of science might reasonably dispute some of these precepts, which are crumbling under the onslaught of quantum mechanics and complexity theory, but they still inform the culture and mindset of

---

[1] See Burtt's 1925 treatise *The Metaphysical Foundations of Modern Science* for a thorough discussion of the origins of unquestioned metaphysical assumptions in the conceptual categories established by Isaac Newton.

science. Starting from this implicit metaphysics, consider these other ways that science resembles religion.[2] Science has:

- A procedure for attaining Truth (the Scientific Method)

- Elaborate divinatory rituals to gain knowledge (experiments)

- Further rituals (technology) by which we manipulate reality

- Invisible universal spirits (such as "energy" and "forces") that are responsible for all movement and change

- An esoteric language understandable only by initiates

- Teachings on human nature

- A creation story (the Big Bang and Darwinian evolution)

- Invisible entities (like electrons, mitochondria, etc.) that can be revealed with the help of special implements (like microscopes)

- Special rituals for the purposes of healing (medicine)

- A priesthood, a laity of various degrees of piety, and infidels

- Training for and initiation into the priesthood (graduate school)

- Orders and associations for the priests

- "Preachers"—science writers and popularizers to bring the gospel to the lay masses

- Legendary saints and héros (Darwin, Newton, Archimedes, Einstein, Maxwell, Bohr...)

- Martyrs for the cause (Bruno, Galileo)

- Mainstream sects and wacky cults

- Extremists, fundamentalists, and tolerant moderates

- Doctrinal schisms, heretics, and apostates

---

2   Not all religions have the following traits, but most institutional religions bear many of them. Obviously, the closest cousin to science in its structure is Catholicism.

- Excommunication of heretics (cutoff of funding, blacklisting from journals)

- A system of ethics and morals (e.g., rational choices, scientific policies)

- A system for the indoctrination of youth

The point here is not to dismiss science on the grounds that it is, after all, nothing but a religion. To do so would be to commit a subtle error: adopting science's own conception of religion as a term of critique. If, however, we reject the implicit devaluing of religion that comes from contradistinguishing it from science-as-the-royal-road-to-truth, then to name science as a religion is no longer to disparage it. Instead it opens up new questions. We might ask "What are the limitations of the kinds of technology that are available from within this worldview?" and "What other religions—systems of metaphysics, perception, and technology—might be born of the current crisis and needed to address it?" We also might inquire as to what science might become if we abandon some of its metaphysical assumptions. What does it become when we recognize that observer and observed are inextricably entwined? When we recognize the consciousness and agency of all matter? When we cease privileging quantitative over qualitative reasoning?

Science is not alone among religions in having a shroud of dogma and institutional dysfunction around a core spiritual truth. The spiritual essence of the religion of science is the opposite of its institutional arrogance: the Scientific Method embodies a deep and beautiful humility. It says, "I do not know, so I shall ask." When science is healthy, that humility takes form as critical thinking, patient empirical observation, hypothesis testing, and perhaps most importantly, communities of knowledge seekers who criticize, refine, and build upon each other's work. The true scientist is always open to being wrong, even at the cost of funding, prestige, and self-image.

Held by a culture of practice, these qualities of humility and experience over time are what make a path of knowledge into a science. My call here is therefore not to discard science but to expand it, to include what it has ignored.

Ecofeminists and deep ecologists have critiqued science for its propensity to abstract, isolate, and distance the observer from the beingness of the observed; to render the world into an object. Francis Bacon conceived the experimental method as an interrogation of nature, even a rape of nature, forcibly penetrating to her deepest mysteries. How might it change if we conceive it as a conversation, not an interrogation; a lovemaking and not a rape? What if we saw science not as a means to force nature into our categories, but as a way to expand the reach of our senses in order to better behold the beloved?

I raise these questions with a certain trepidation, since the conventional view is that any rejection of science is a step backward into long-discredited myths, irrationality, and superstition. I don't want to be lumped in with the ignoramuses. It seems quite evident to most people that the problem today isn't too much trust in science but, to the contrary, too little. Consequently, you might think that even if my previous points are, perhaps, philosophically valid, to bring them up in the context of climate change is a strategic error that will embolden climate change deniers and give cover to polluters. I raise them nonetheless, because both the metaphysical assumptions of science and its institutional expression are part and parcel of the system that has laid waste to the world. Science's reduction of reality to number mirrors the conversion of nature to money. Its universalization of matter into generic particles mirrors the standardization of people and commodities in the industrial economy. And the technology that comes from science facilitates both.

Though it is evolving, science as we have known it (and still to a great degree) has trained us:

To see the world as a bunch of insentient things

To make decisions "rationally"; that is, based on utilitarian calculations

To see the observer as independent from the observed

To see nature as an object of manipulation and control

To ignore the immeasurable and qualitative (spirit, beauty, sacredness, etc.)

To think in mechanistic rather than organic terms

The anti-science public may, for all its apparent ignorance, be tapping in to an authentic intuition about the limits of science as a decision-making compass and the ultimate arbiter of truth. We have to stop seeing public rejection of science and authority in general as some kind of troublesome insubordination, but rather look for the uncomfortable truth within it.

When we say, "Trust the scientific consensus on climate change," we are also implying:

Trust the social processes by which this consensus is formed.

Trust the other things about which a scientific consensus is declared.

Trust the basic approach to knowledge that science represents.

Trust the tacit metaphysical and ontological assumptions that underpin science.

Trust other institutions that draw their legitimacy from science.

Trust the power of scientific technologies to solve our problems.

In various ways, all of these things we have trusted have contributed, and continue to contribute, to the ongoing devastation of the biosphere. This presents the more radical environmentalist with a conundrum when invoking science in the fight against climate change, because it requires a buy-in to the very same systems of intellectual authority that have long presided over and defended our ecocidal system. In addition, the urgency of action invites the further empowerment of existing institutions, which are the only ones capable of immediate action. The climate activist finds herself in the uncomfortable position of championing and fighting the establishment at the same time.

We are on shaky ground indeed if we are to rely on trust in the institution of science, and by extension trust in authority generally, to deliver us from climate catastrophe. On the rhetorical and strategic level, we need to reach beyond the good little schoolboys and schoolgirls who trust science and believe what the teacher tells them is important. And we need to cleanse ourselves of the stink of self-righteousness that comes when we hold in contempt those who don't understand science (or patronize them as recalcitrant rubes to be "educated" in a dumbed-down version of it). "Science says" is not going to reach the farmers, hunters, ranchers, and other people who (in the United States) typically have conservative political identities, voted for Donald Trump, and are polarized into the climate skeptic position. Nor will it much impress working-class people who feel, quite understandably, that the establishment has betrayed them.

Recent political events like Brexit and the Trump election point to a growing popular rejection of established authority. Normally attributed to bigotry, xenophobia, and, tellingly, "irrationality," these events point to a morbid crisis in the legitimacy of our dominant institutions and the elites who run them. It will only worsen as the concentration of wealth intensifies and the social contract frays further. It will worsen as the mainstays of our society—medicine, education, and law—veer closer to absurdity in their dysfunction. It will worsen as progressive and conservative governments alike fail to fix the political system. It will worsen as people realize the power and beauty of what has germinated outside the bounds of normalcy, in the "alternative" realms.

Many people have direct experiences that contradict what science and authority generally tell them is real and possible. A friend's lifelong menstrual cramps disappear for good after a few acupuncture sessions, in spite of her extreme skepticism. A woman recovers from "incurable" stage four pancreatic cancer. A man experiences direct communication with his ancestors in an iboga ceremony and ends his drug addiction. Rival gangs meet in a restorative circle and come to peace. My son's teenage friends see a UFO. Experiences like these open people to further experiences. When the "impossible" happens, we begin to question the bounds of the conventionally possible.

Some of the most highly educated people I know are devoted to astrology: an academic philosopher, a law professor, a medical anthropologist. These are not people who are too stupid to understand that the gravitational influence of other planets is negligible. Nor are they ignorant of confirmation bias, nor of the mind's proclivity to perceive illusory patterns in random noise. These are highly intelligent, self-reflective people. You can write them off as superstitious nincompoops who are less rational than you are, but on what evidence? Because you just *know* that the worldview of institutional science, and its account of causation, is identical to reality itself? Likewise, can you be sure that cultures around the world that have maintained divinatory practices for thousands of years were simply oblivious to the mind's capacity for self-deception? Is it that we are wise and they are stupid, that we are advanced and they are primitive, and that our historical duty is to replace their inferior ways of knowing with our superior ones? That mentality seems more like part of the problem than part of the solution.

Ironically, many of the very same people who embrace energy healing, astrology, crop circles, and so on also add their voices to the call for "science-based policy" on climate. Meanwhile in their own lives, they apply *I Ching*–based policy, Tarot-based policy, or astrology-based policy. This exemplifies the wall of separation that keeps spirituality and politics in separate realms. That is a division that must crumble. The key to our salvation lies beyond what science currently offers—it lies in facing the world as a living being, a sacred being, and a beloved being. From that place, technologies and practices emerge far beyond what science thinks is possible today. The astonishing results of regenerative agriculture are just a taste of what can happen when we think, "Land, I know you want to heal. Please tell me how to serve you. Land, I know you want to give. Please tell me how to serve you. Land, I know you want to fulfill your highest purpose. Please tell me how to serve you."

This is the state of heart and mind from which the insights of regenerative agriculture and ecological healing arise.

Science can be a powerful tool to ask these questions and hear the answers. I am not advocating replacing it with Tarot decks, or indeed with the divinatory practices of other cultures who did indeed use exquisite rituals to maintain balance with the land. What needs to change is the impulse behind the science: the manipulation of a world it sees as dead—atoms and void. When that view changes, science will morph into something we can hardly recognize. It will share the animating force of indigenous ways of communicating with nature; it will be a step toward recovering our own indigeneity. That word must mean to be truly of a place, to be intimate with a place and all its beings. In the end, it does not matter if we enact the technological rituals of science, or some other religion. What matters is that we return to love.

Look at it this way. You don't enact "science-based policy" in affairs of the heart, do you? You enact, I hope, love-based policy, or maybe hate-based policy or fear-based policy. You might dress them up with reasons, but love is not reasonable. If we want to enact unreasonable commitment to the healing of the earth, we need to make our relationship with it into an affair of the heart. Otherwise, the catastrophists may be right.

If We Knew She Could Feel

It is not only Western climate catastrophists who warn of a great dying to come. Many indigenous people also see a great peril upon us. Their warning does not invoke rising levels of greenhouse gases; it references another matrix of causality that involves the desecration of life itself. This deeper causal system also suggests a deeper set of responses, all of which come down to holding life and matter sacred again. It offers new hope, an exit from the futility of the endless "fight" against climate change.

It won't be surprising to anyone who has read chapters 4 and 5 of this book that many of their warnings are on one level simply about ecosystem destruction. Here are the words of the remarkable Yanomami shaman Davi Kopenawa from the book *The Falling Sky:*

The forest is alive. It can only die if the white people persist in destroying it. If they succeed, the rivers will disappear underground, the soil will crumble, the trees will shrivel up, and the stones will crack in the heat. The dried-up earth will become empty and silent. The xapiri spirits who come down from the mountains to play on their mirrors in the forest will escape far away. Their shaman fathers will no longer be able to call them and make them dance to protect us. They will be powerless to repel the epidemic fumes which devour us. They will no longer be able to hold back the evil beings who will turn the forest to chaos. We will die one after the other, the white people as well as us. All the shamans will finally perish. Then, if none of them survive to hold it up, the sky will fall.[3]

Here Kopenawa expresses a belief widespread among indigenous people: that human activity, including ritual activity, is part of the glue that holds the world together. When we forget our proper function and cease to serve life, the world falls apart.

The tribes of the Colombian Sierra Nevada de Santa Marta (of whom the best known are the Kogi) have a similar belief.[4] They believe that a black line, a network of hidden connections, links all the sacred places on Earth. If that line should be broken, calamities will ensue, and this beautiful world shall perish. Destroying a forest here, draining a swamp there, might have dire consequences across the globe. The shamans cannot perform their work of maintaining the balance of nature much longer in the face of our depredations.

How are we to interpret these warnings?

Several interpretations occur to the Western mind, all unsatisfactory. Most of us would no longer be so crude as to dismiss the warnings as the magico-religious rantings of benighted primitives whom we must awaken from their silly superstitions. Today we have more sophisticated ways to deafen ourselves to their message.

The first we might call "ontological imperialism." It would be to say, "Yes, the indigenous are onto something after all. The black line is a metaphor for

---

[3] From the preface to Kopenawa and Albert (2013).

[4] The other tribes are the Arhuaco (or Ika), Wiwa, and Kankuamo. Of these, the Kankuamo are largely assimilated. The Arhuaco are politically active in the movement to protect indigenous rights, while the Kogi have retreated up the mountain to minimize contact and protect their culture. All have suffered at the hands of coca growers, paramilitaries, land developers, and so forth. In this section I will usually refer to the Kogi, although much of what I say is also true of the other tribes.

ecological interconnectedness. "*Xapiri* spirits" is code for the hydrological cycle. The indigenous are keen observers of nature and have articulated scientific truths in their own cultural language." That sounds fair enough, doesn't it? It gives them credit for being astute observers of nature. However, this view takes for granted that basal reality is that of scientific materialism, thereby disallowing the conceptual categories and causal understandings of the indigenous. It says that fundamentally, we understand the nature of reality better than they do.

If their message were merely "We must take better care of nature," then the above understanding would be sufficient. But people like Davi Kopenawa and the Sierra Nevada tribes are inviting us into a much deeper change than that. Do we understand the nature of reality better than they do? It once seemed so, but today the spawn of our supposed understanding—social and ecological crisis—gnaw at our surety.

A second and related form of deafness is what Edward Said called "Orientalism"—the distortion (romanticizing, demonizing, exaggerating, reducing) of another culture to conform it to a comfortable and self-serving narrative. Accordingly, we could turn the Kogi into a kind of cultural or spiritual fetish object, subsuming them into our own cultural mythology, perhaps by making them into an academic subject and stuffing their beliefs and way of life into various ethnographic categories. In that way we make them safe, we make them ours. It is another kind of imperialism.

We might do the same by inserting their messages into a comfortable silo called "indigenous wisdom," elevating the indigenous to superhuman status—and, in the process, dehumanizing them as well. It is not true respect to worship an image—the reverse image of our own shadow—that we project onto another culture. Real respect seeks to understand someone on their own terms.

The Sierra Nevada tribes are famous today thanks to two films, *From the Heart of the World* and *Aluna*.[5] I have always been a little uncomfortable

---

[5] This section was adapted from a review of *Aluna* that I wrote for *Tikkun* magazine: "*Aluna*: A Message to Little Brother" (Eisenstein, 2015a).

with documentaries about other cultures, because they of necessity objectify their subjects, turning them into the material of a (video) "document." By documenting them, we incorporate them into our world, into a safe educational or entertainment or inspirational frame, and into the Debordian society of the spectacle. Fortunately, these films are not documentaries.

Who is the filmmaker here? Ordinarily one would say it was Alan Ereira, a former BBC producer who came in with his cameras and crew. But that's not what Ereira says, and that's not what the Kogi say either. According to them, the elders noticed the accelerating degradation of the planet and contacted the outside world to deliver a message that we must stop the destruction. They did so first in the early 1990s with *From the Heart of the World*, after which they again withdrew from contact.

Obviously, we didn't heed their message. "We must not have spoken it clearly enough," they concluded, and so they sought out Ereira again to make a sequel. The cynical observer, practiced with the tools of postcolonial analysis, might think that the assertion that "the Kogi have requested this film be made in order to convey their message" is a mere cinematic trope, or a way to preempt charges of exoticism, Orientalism, and cultural appropriation. However, that analysis is itself a kind of colonialism that sees the Kogi patronizingly as the helpless pawns of the filmmaker, and discounts their own explicit assertion that they have called the filmmaker back in order to transmit an important message to "little brother" (us).

Dare we take the Sierra Nevada elders at face value? Dare we hold them in full agency as authors not only of this film, but of a message sent to us on their initiative? To do so reverses the power relations implicit in even the most post-colonially sensitive ethnography, in which the distinction between the ethnographic subject and the ethnographer is usually preserved in some form (and institutionalized when, with all due disclaimers, it appears in academic publications). Anthropologists don't normally grant ethnographic populations agency as the originators of messages to academia.

In these films, the colonial gaze is turned back on us: sternly, imploringly, and with great love. The elders tell us, "You mutilate the world because you don't remember the Great Mother. If you don't stop, the world will die." Please believe us, they say. You must stop doing this. "Do you think we say these words for the sake of talking? We are speaking the truth."

Why hasn't little brother listened? It has been nearly thirty years since the Kogi elders first spoke their message to the modern world. Perhaps we have not listened because we have not yet come to humility. We continue to try to somehow box, contain, and reduce the Kogi and their message so that it can rest comfortably in our existing Story of the World.

In this book I have proposed that our reductive Story of the World underlies the world's literal reduction: extinction, soil impoverishment, ecosystem collapse, etc. The Kogi offer a similar teaching. They say that thought is the scaffolding of matter; that without thought, nothing could exist. (This is not an anthropocentric view, because they do not consider thought to be merely a product of the human mind. Thought is prior to human beings; our minds are but one of its receivers.) The official *Aluna* website describes the Kogi's view thusly: "We are not just plundering the world, we are dumbing it down, destroying both the physical structure and the thought underpinning existence."

Thankfully, the requisite humility to truly hear the Sierra Nevada elders is fast upon us, born of—what else?—humiliation. As our cultural mythology falls apart, we face repeated humiliation in the failure of our cherished systems of technology, politics, law, medicine, education, and more. Only with increasingly strenuous and willful ignorance can we deny that the grand project of civilization has reached a dead end. We see now that what we do to nature we do to ourselves; that its plunder brings our poverty. The utopian mirage of the technologist and the social engineer recedes ever further into the distance.

The breakdown of our categories and narratives, the breakdown of our Story of the World, gives us the gift of humility. That is the only thing that can open us to receive the teachings of indigenous people—to truly receive them, and not merely insert them into some comfortable silo called "indigenous wisdom," as if they were a museum piece or a spiritual acquisition.

I am not suggesting that we adopt, part and parcel, indigenous cosmology. We need not imitate their shamanic practices or learn to listen to bubbles in the water. What we must do is embrace the core understanding that motivates the attempt to listen to water in the first place: the understanding that nature is alive and intelligent. Then we will find our own ways of listening.

The Western civilized mind does not easily comprehend the idea of the intelligence of nature except through anthropomorphizing or deifying it—another attempt at conquest.

Granting subjectivity and agency to nature and everything in it does not mean to grant *human* subjectivity and *human* agency, making them into storybook versions of us. It means asking "What does the land want?" "What does the river want?" "What does the planet want?"—questions that seem crazy from the perspective of nature-as-thing.

Materialism, however, isn't what it used to be. Science is evolving, recognizing that nature is composed of interdependent systems within systems within systems, just as a human body is; that soil mycorrhizal networks are as complex as brain tissue; that water can carry information and structure; that the earth and even the sun maintain homeostatic balance just as a body does. We are learning that order, complexity, and organization are fundamental properties of matter, mediated through physical processes that we recognize—and perhaps by others we do not. The excluded spirit is coming back to matter, not from without but from within.

So the question "What does nature want?" does not depend for its coherency on anything supernatural, no external intelligence. The wanting is an organic process, an entelechy born of relationship, a movement toward an unfolding wholeness.

In that understanding, we can no longer cut down forests and drain swamps, dam rivers and fragment ecosystems with roads, dig pit mines and drill gas wells, with impunity. The Kogi say that to do so damages the whole body of nature, just as if you cut off a person's limb or removed an organ. The well-being of all depends on the well-being of each. We cannot cut down one forest here and plant another there, assuring ourselves through the calculus of net CO_2 that we have done no damage. How do we know we have not removed an organ? How do we know we have not destroyed what the Kogi call an *esuana*—a key node on the black thread scaffolding the natural world? How do we know we have not destroyed a sacred tree, what the Kogi call "the father of the species," upon which the whole species depends?

Until we can know it, we'd best refrain from committing further ecocide on any scale. Each intact estuary, river, forest, and wetlands that remains to us, we must treat as sacred, while restoring whatever we can. Davi Kopenawa and the Sierra Nevada elders agree: we are close to the dying of the world. This warning does not contradict the possibility I raised in chapter 7, that humanity might yet survive in a ruined world, a concrete world, a dead world. The world might die and yet we might live.

Science is beginning to recognize what many cultures have always known. An invisible web of causality does indeed connect every place on Earth. Building a road that cuts off the natural water flow at a key site might initiate a cascade of changes—more evaporation, salinization, vegetation die-off, flooding, drought—that have far-reaching effects. We must understand that as exemplifying a general principle of interconnectedness and aliveness. Otherwise, we are left only with the logic of instrumental utilitarianism as reason to protect nature—save the rainforest because of its value to us. But that mindset is part of the problem. We need more love, not more self-interest. We know it is wrong to exploit another person for our own gain, because another person is a full subject with her own feelings, desires, pain, and joy. If we knew that nature too were a full subject, we would stop ravaging her as well. As an elder says in *Aluna*, "If you knew she could feel, you would stop."

If you knew she could feel, you would stop. Isn't it obvious too, that as long as we don't know she can feel, we will never stop? Isn't it obvious that we need a Story of the World that helps us to know that she can feel?

The Powers of the Land

The problem with the mechanistic view of nature-as-thing is not only that it numbs our compassion and facilitates our plunder. It also cripples our ability to serve as agents of positive transformation. One reason is that from the mechanistic perspective we cannot fully understand the needs of land, ocean, soil, water, or forest, just as I could not fully serve my son's needs if I saw him as a biomechanical robot merely requiring precise inputs of various substances.

Another reason is that it leaves us without allies. If the world outside ourselves lacks purpose, intelligence, and agency, then change is entirely up to ourselves, dependent on how much force we can exert on matter.

Without allies in the cause of ecological healing, the situation is grim. As I asked before, can we beat the military-industrial-financial-agricultural-pharmaceutical-NGO-educational-political complex at its own game—the game of one force against another? If we have no allies, if humans are the sole possessors of intention in a random world, then we are lost.

What becomes possible when we believe we have companions inconceivably more powerful than ourselves with whom we can align? What becomes possible when we seek to participate in a larger ordering intelligence?

Older cultures commonly believed that mountains, rivers, animal species, the ancestors, and other seen and unseen beings participated in human affairs and could alter the course of history. Can we access these beings as allies too?

A word of caution here: this is not the kind of alliance that we are accustomed to in war thinking. We are not enlisting an even bigger force in a contest of force versus force. In fact, the allies abandon us when we inhabit that mentality, which is kindred to the story of nature-as-thing. That mentality casts us into a universe in which the allies do not exist. They become invisible to us through the lens of instrumental utilitarianism, through the lens of "resources," minerals, commodities, and profit. One might ask, "If the powers of nature are so great, then why haven't they put an end to the destruction?" If there is, as indigenous people tell us, a power in the land, in the mountains, in the forest, in the waters, that is greater than human power, why are all of these dying at human hands? It is because theirs is not a power of force versus force.

Stephen Jenkinson puts this poignantly in his magnificent book *Come of Age:*

> *The wild doesn't play by the rules as we know them. But the wild is governed by a kind of ruthless etiquette; it does not preserve itself by subverting its wild soul. Who among us has not been on some camping trip or the like and seen the inexorable creep of civilization coming on, the intrusion of distant neon on what was once a dark night? How many times have you heard older people remember a time, not so long gone, when every developed part of your life was a field of Queen Anne's lace and deer spoor? How many of you have harbored a wish that something of the wild would rise up and smite Big Pharma, or Big Agra, or the Military Industrial Complex, or your local bad guy equivalents, Armageddon-movie style, just enough to reinstate the boundary line and lend us a little hope, and then forgive the rest of us enough to allow a little conscious ecotourism to help with the maintenance costs? The wild seems terribly vulnerable in our time. To respond in kind to the indignities and rapacious practices we oblige it to endure would be to practice our kind of "desolation by payback," our kind of retributory justice—the very undoing of the wild's other-than-human ways of being itself. So this defenselessness sustains the wild's soul, you could say. It is heartbreaking. And if the wild expires at our hands in decades to come, species by species, place by place, it does so as the wild does, not in soullessness, not in punition, but in silence.[6]*

Yet this does not mean we cannot come into relationship with these beyond-human powers. We cannot impose our terms upon them or pervert them to our ends, but we might align ourselves with the tiny sliver of theirs with which we intersect.

The Australian activist Daniel Schneider told me a story about a protest against a fracking project in New South Wales. Thousands of people, including many aborigines, occupied the site, set up a camp for three months, blockaded the roads, chained themselves to cars, and sat on top of poles to block heavy equipment from entering. "Basically, we were

---

[6] Jenkinson (2018). From a prepublication draft sent to me by the author.

preparing for battle," Dan said. They found out that a force of eight hundred police was being prepared to move in the next week, along with agents provocateurs to create a pretext for mass arrests. The protesters prepared for a showdown—not to actually fight the police of course, but to battle for public attention and media exposure. They had drones and cellphone cameras sending live feeds to global activists. They were prepared to win the war of public perception and expose the villainy of the police and the government.

As tensions were reaching their peak, Dan proposed an idea to a group of aborigines at the site. Everyone felt the foreboding that they were entering a losing battle, so why not try something else? Since they knew media helicopters were coming, why not make giant art installations visible from the air for them to film, instead of the usual script of police arresting activist hippies? The aborigines loved the idea, brought out their dreaming stories, and soon had sketched designs for two-hundred-foot giant rainbow serpents and other figures to be drawn on the ground with sacred ochre. They also planned to greet the police ceremonially, with giant fires making sacred eucalyptus smoke, and five hundred men painted in ceremonial colors with clapping sticks and didgeridoos.

The next morning Dan got a phone call. The government had canceled the fracking license.

Later an indigenous elder woman came to him. "Thank goodness we let go of conflict. That is why we were successful. I've seen it before," she said. "Usually it is the same story. The police come in, all the blackfellas get arrested, lots of the whitefellas get arrested, and the project continues. But this time, because we let go of conflict and entered into art and ceremony, the ancestors of the land could come in and exercise their power."

I heard a different version of this story from my dear friend Helena Norberg-Hodge, who lived an hour from the site. According to her, the victory came thanks to the "knitting ladies"—older women, white and aboriginal, who, as they quietly went about their knitting, kept peace in the encampment, restrained the fighting and drunkenness that broke out among the men, and opened up backdoor communications channels with the police. Working behind the scenes, they shifted the dynamic away from confrontation; furthermore, by humanizing the protesters they disabled the narrative of "environmental extremists" that would have facilitated a police invasion.

I see these two versions of the story as complementary, not contradictory. How do the "ancestors of the land" exercise their power, if not through the quiet faith of the knitting ladies? What power sustains those women? What holds them in peacefulness against the onslaught of us-versus-them thinking?

In the story of force versus force, the deeper sponsoring assumption is that if anything purposeful is to happen, we have to *make* it happen. It has no room for the agency of other beings to engineer synchronicity. It locks us in a world where the ancestors and the powers of the land have no room to operate. Nor does it allow the women to wield their feminine power to hold the peace so that the ancestors and the land can do their work. In fact, it says to them—women, ancestors, and earth powers—"We don't need you. We don't acknowledge you."

This mentality is akin to the geomechanical view of climate change, which also discounts the ability of living systems to maintain conditions for life. They can—if only we let them. But we do not let them; instead we actively degrade and destroy that capacity. We kill forests, lakes, mountains, and swamps in part because we see them already as dead, using the same vision that excludes the powers of the land from operating in ways that are more mysterious than regulating the carbon cycle, the water cycle, and the surface albedo.

By the same token, we must let all these beings "maintain conditions for life" by aiding us in the realms of technology and politics. When we engage in confrontational tactics or fight the destroyers in court, we must remember that it is not all up to us. We must remember that purposive change is possible beyond what we direct ourselves. We must remember that this is not a fight we can win just by fighting.

Have you ever noticed in life that the most striking synchronicities seem to happen in times of uncertainty? When one moves to a new city without a plan, or travels without an itinerary, or does something out of the ordinary with no idea of what will happen, then quite often an amazing (sometimes life-changing) meeting or stroke of luck or "chance" encounter occurs. They rarely happen when everything is planned, predictable, and controlled. It is as if the spirits have no room to come in.

To enter the realm of synchronicity, the aid of the ancestors, and alliance with the powers of the land is not the same as sitting around doing nothing and wishing it will happen. It is not enough to "send positive energy." A sacrifice of some sort is required, something that involves risk or loss. It might be the sacrifice of time, energy, and money. It could be a sacrifice of certainty or control, an act that feels like a step into the true unknown. It could be a demonstration of commitment that feels real to you. It might be the sacrifice of "winning"—of having the satisfaction of seeing your opponent admit he was wrong. It might be to sacrifice setting up the situation so that you get to be the leader or get credit for the success. It might be the sacrifice of a polarized, dehumanizing view of the other side that makes you out to be the good guy. It might be the sacrifice of a self-image; for instance, being the one with the answers.

Another way to understand the necessity of a sacrifice is that who I am right now is not in full alignment with the more beautiful world I wish to help create. In order to be its effective servant, and in order to inhabit the reality in which it is possible, I must undergo a transformation. Something will be lost and something will be gained. I must give something up to align with the future that calls to me.

The sacrifice I speak of is usually not deliberate, but the result of a realignment of the self to a different life purpose or creative goal. What is deliberate is to commit life energy in service of the prayer, to take action in the 3D world. Conventional actions, especially those demanding hard work, significant money, or risk of imprisonment, constitute a commitment ritual that communicates to the unconscious and to all who are watching, "I am serious about this."

That which hears our prayers gets fed up with prayers that aren't serious. Often in our culture, we wish for things to be one way but act in direct contradiction to that wish. So the Listener wonders, "Do you really mean it? Let me make sure." The Listener then creates a situation—a challenge or a setback—that gives the wisher a chance to clarify whether she really means it.

The environmentalist Mark Dubois told me a story about a campaign he and other environmentalists waged in the 1970s and early '80s to stop the New Melones Dam from being built on a pristine stretch of the Stanis-laus River. Their group of activists tried everything from legal challenges to petitions to lobbying for legislation to physical direct action (Mark chained himself to a boulder to prevent the authorities from filling the reservoir), all to no avail. They poured so much of their hearts and souls into the campaign that when they finally lost, their pain and grief were so great that many could not bear to visit the flooded canyon. It felt like a total defeat. Yet, the New Melones project marked a turning point. It was the last dam of its size built in the United States; since then, every new dam project has met with stiff opposition, and more dams have been removed than have been constructed.

Certainly, one could cite mundane explanations for the end of the dam-building era. Few viable sites remain in North America; the visibility and cost of the New Melones fight made the authorities lose their appetite for further projects; the resistance heightened public awareness of the damage dams cause. All true, yet on another level we might understand the failed campaign as a kind of prayer. When we put everything we've got into the service of a vision, the world takes notice and reality shifts. Our failures are our prayers. This is not to suggest we commit to an impossible cause, hoping that performing the rituals of protest will magically bring the impossible result we wish for. It means doing the best we can based on the knowledge we have, knowing that our sincere commitment will impact the world. No sincere action is ever in vain.

We cannot be sure our prayers will be answered in the form we expect. We can be confident though that our prayers are at least heard. We are not alone here. Something is watching. Something is listening.

I can imagine my evangelical Christian friends saying, "Yes, that 'something' you are talking about is God." I agree with them, except that they conceive God as an immaterial being, a spirit that directs matter but is separate from it. Holding matter itself as insensate, they agree with scientific reductionism. I would say that the "something" that is listening is everything: earth, sky, water, air, rocks, trees, animals, plants ... along with beings we do not see and that have no name (in English, anyway). Matter is sentient, watching, listening; God, you might say, is in all things, and nothing is not God.

Reanimating Reality

The more closely we participate in the affairs of earth, sky, soil, rocks, and so on, the easier it is to see the God in all things. This is not a perception exclusive to animistic cultures. The poet David Whyte recounts a visit with a Scottish fisherman on a remote island, who lived the traditional ways. He said a prayer for every significant act of the day: a prayer for getting out of bed, a prayer for drawing the curtains, a prayer for breaking bread, a prayer for getting into his boat, a prayer for casting the net. His was a world thick with being. Something is always watching, always listening. He was never alone, because the whole world was alive.

The reanimation of our world is crucial to ecological healing. If we live in the perception that the world is dead, we will inevitably kill what is alive.

How does reanimation happen? You might hold highly developed philosophies of nondual spirituality, animism, pantheism, or panentheism, but act automatically from the old story when push comes to shove. Our entire cultural conditioning militates against the deep trust that comes from knowing that God sees everything, from knowing that all beings are alive and listening, from knowing that every action has cosmic significance. To mentally embrace a new story is a first step, but alone it is insufficient to undo generations of cultural programming.

Go ahead and try talking to a tree or a pond. If you are like me, a voice in your head will hector you: "It isn't really listening. It can't understand you. You are being silly." And even if it seems like the tree is talking back to you, do you wonder if maybe it is just your imagination? Normally, people need some help to inhabit the Story of Interbeing deeply enough to consistently act on it.

The help comes in the form of a direct experience. We can't force the other beings of this world to reveal themselves in their beingness, but we can ask them. The way to ask is to give attention to your longing: your longing to rejoin the living universe, to be companioned.

I will give an example of such help that I received a couple years ago, when I visited Taiwan on my way to hold a retreat in Indonesia. I used to live in Taiwan, and I think a piece of my heart is still there. My old friend Philip picked me up from my 5 a.m. flight and we drove straight into the mountains, nearly an hour on narrow, winding roads, to a spot where he'd heard there was a sacred grove of trees. We parked in a lot near the trailhead and hiked up a steep, narrow trail that required the use of ropes in some places. But after a couple hours of hiking, we still hadn't found the grove, and we needed to turn back soon because we were tired and I was supposed to speak at 3 p.m. in Taipei. When we came to a nasty muddy uphill section, we considered turning around.

"Let's go a little further," I suggested, "up this hill. Maybe we'll be able to see it from there." We got past the muddy section and there wasn't a view there, just another rise, but then we saw a little sign that said, "Come on! The sacred grove is only 5 minutes away!"

It was as if the sign were written just for us.

Soon we arrived at the grove. The trees were incredible. Two-thousand-year-old massive trees, trunks fifteen feet or more in diameter, with ancient branches thicker than I am tall, covered with ferns and other plants, each a whole ecosystem unto itself. It was impossible to look at them without a nearly overwhelming feeling of being in the presence of a divine being. We were awestruck. Neither of us spoke for quite some time.

I thought about how the whole forest used to consist of trees like these. They'd all been cut down except these seven or eight grandfathers, spread over maybe an acre of woods. I wondered if the trees were angry that humans had cut down all their companions. "Do you think the trees are mad at us?" I asked Philip.

He knew exactly what I meant and took the question seriously. After a while he said, "No. They are happy that we are here." His words rang with truth.

Later I understood why the trees were happy. They were happy that I had even asked the question and that Philip had taken it seriously. Because that question came from a place of actually seeing the trees as real beings that could be angry or sad, instead of seeing them the way the timber companies had seen them, as the mere stuff of profit, or as most hikers saw them, as mere spectacles to photograph. Dear reader, have you ever had the experience of finally being seen for who you are? Women and black people especially know what it is to be seen as less than a full being, but even my fellow white straight males know what it is to be a mark, a sales target. So I think the trees were happy that we humans were rejoining them in the community of all being.

As we descended the trail back to the car, something odd happened—a subtle shift in reality, as if we had entered into a dream world where everything took on a symbolic resonance. A troupe of monkeys visited us, swinging directly over our heads. When we got to the parking lot Philip said, "I'm a little worried about the keys. They don't seem to be in my pocket."

We looked everywhere in his backpack and on the ground. Finally I looked in the car and there they were, stark and taunting on the front seat. The car was locked and the windows were up. It was like a dream: "The key to the vehicle is locked in the vehicle itself." I suppose there must be a spiritual teaching there.

My friend got anxious, and the more anxious he got the more relaxed I became, wondering what adventure the universe (or the trees) had arranged for us. I was certain we would somehow make it on time, without being concerned at all whether we would. Everything felt perfect.

This was a remote spot—the nearest village was twenty minutes (driving) away. Philip got out his cellphone to call someone to come get us. Of course the battery was dead. There was a little house nearby and we asked the guy there if he had any tools to break into the car. No. How about a garage or locksmith? He let us use his phone but didn't have the number of anything useful except the nearest police station. I called them and they said they'd send someone.

An hour later a police car came. The police were businesslike and gruff at first, but only because they were embarrassed that they had no clue how to get into the car either. They called a garage but when they were told the price, they were indignant on our behalf and told the garage not to send anyone. So there we were, four dudes without a clue, shuffling our feet. Finally one of them said, "You're just gonna have to break a window."

This is like a dream, remember? So how many blows with a rock did it take to overcome my inhibitions and break the window? Three blows. And a little shard of glass flew out and drew a single drop of blood from my finger.

Trying not to be late, we were driving really fast but suddenly a fruit stand called out to us and it turned out to be selling native varieties of apples and tangerines, little Taiwanese ones less than two inches in diameter, organic, and amazingly flavorful, as if all the flavor of a big one were packed into each. Everything was perfect. I shared my sentiment with Philip, who, facing the prospect of returning the car to the friend he'd borrowed it from, was understandably less enthusiastic about our day so far, although touched by the magic of the trees as much as I was. Wryly he asked me, "Is there anything else I can do to make the day even more fulfilling for you?"

"Well," I said half-jokingly, "you know those 'earth guavas'?" (These are tiny guavas that grow in Taiwan and are almost never sold commercially.) "I'd sure like some of those."

"I don't know if I can help you there."

We showed up to the venue exactly on time and I gave my speech splattered with mud. Just before I began, an old friend whom I hadn't seen in twenty years came up and gave me a bag. "I thought you might like these," he said. In it were boiled peanuts, *lingjiao*, and, you guessed it, earth guavas.

It was like the island of Taiwan was saying, "You don't believe me that I love you? Well, here are some earth guavas just to make sure."

Of course it could all be coincidence, but it seemed like the trees gave me that gift, using my friend as their instrument. "Dang," I joked to myself, "if I'd known any wish would be fulfilled, maybe I would have asked for more than guavas."

I offer you this story to suggest that when we enter the world-story in which all beings are sentient, the world comes alive with them. We begin to experience synchronicities that confirm the universe is intelligent. Or is it that we are just noticing them more? The mind of separation wants proof to precede belief, but I find it is often the other way around. Thus we face a choice. Which world shall we live in? It echoes the choice posed in chapter 7: a concrete planet, or a planet profuse with life? A beautiful world or an ugly world? A living world or a dead world?

If we want a living world, we have to act from the place where the world is alive.

12

Bridge to a Living World

Knowing the depth of the initiation upon us, where do we go from here in practical terms? Without a bridge from the realm of metaphysics to the world of policy, we risk making the Story of Interbeing into a mere philosophy.

Such a bridge must span a great gulf. Notions like "the voice of the land" seem positively ridiculous in the context of current public policy conversations.

On one side of the gulf are the solutions that are necessary for the timely healing of Earth. Visiting that territory, one feels tremendous hope. Earlier this year I visited the Occidental Arts & Ecology Center, headed by the brilliant and loquacious Brock Dolman, where I saw first-hand the practices that could quickly reverse the course of global ecocide. The multistoried agroforestry, the native species restoration, the water retention features, the composting toilets ... none of these are unrealistic fantasies. There they were, in front of my face. At that moment, I knew that my heart's foretelling of a more beautiful world coincides with an actionable reality.

On the other side of the gulf are dominant practices, policies, and perceptions. Real as a more beautiful world seems while I'm at the OAEC, inevitable as it seems when I tune in to Brock's knowledge and intelligence, the OAEC is after all the fringe of the fringe. Permaculture principles practiced there for thirty years are not even a twinkling in the eye of the Secretary of Agriculture. Its budget is less than 0.001 percent of the budget of the agricultural-industrial complex. As this book has argued, most of the responses visible on the progressive radar—such as commercial organic agriculture and renewable energy—are still deeply conventional, freighted with beliefs and practices that contribute to the problem.

Nonetheless, the gulf between these two words is shrinking, thanks to the movement of the tectonic plates—the myths, values, and unconscious agreements—of our civilization. As they shift, formerly unrealistic proposals drift to within a bridge-span of practicality. Yes, the measures I will summarize in this chapter still seem wildly impractical at the present writing; I present them anyway, for three reasons: (1) the collective mind is ready for them to move from the "fringe of the fringe" to the mere fringe—poised to infill the policy vacuum that follows crisis and catastrophe; (2) many of them do not require broad social consensus or institutional endorsement to be practiced on a smaller scale right now by innovators, philanthropists, and landowners; (3) nothing less than these will suffice. Why bow to a "practicality" so narrow that it is tantamount to no change at all?

Here are some of the policies and changes that are necessary over the next couple decades if we are to change course and move toward a living world rather than the Concrete World of chapter 7. Most of them are obvious corollaries to the themes of this book; I will leave two for the end that require some explanation as to why they are so crucial for planetary healing. I am leaving out important reforms like ending mass incarceration or implementing universal basic income that have only indirect (albeit powerful) long-term ecological benefits.

1. Promote land regeneration as a major new category of philanthropy: fund demonstration projects, connect young farmers to

land, and help farms transition to regenerative practices. Provide public funding and government support for this transition as well by shifting agricultural subsidies away from conventional crops.

2. Institute a global moratorium on logging, mining, drilling, and development of all remaining primary forests, wetlands, and other ecosystems.

3. Expand the land protected in wildlife refuges and other reserves. When possible, enlist local and indigenous people in protection efforts to align their livelihood with ecological health.

4. Establish new ocean marine reserves and expand existing ones, with the goal of placing a third to half of all oceans, estuaries, and coastline into no-take/no-drill/no-develop sanctuaries.

5. In the rest of the oceans, establish strict bans on driftnets and bottom trawling.

6. Ban disposable plastic bags for retail purchases. Phase out plastic beverage containers in favor of a refillable bottle infrastructure.

7. Reconstitute the World Bank to serve ecological healing rather than development. Start by declaring the Amazon and Congo rainforests global treasures, purchasing the external debt of countries where the rainforests grow, and canceling the debt at a rate equivalent to the potential income from now-banned logging, mining, and drilling in those areas.

8. Promote afforestation and reforestation projects globally with an emphasis on ecologically appropriate native species.

9. Establish an "eco-corps" to address youth unemployment and restore ecological health by planting trees, building water retention features on public land, deconstructing dams, etc.

10. Change building codes, sanitation codes, and zoning regulations to allow higher density development, tiny homes, composting toilets,

aquaculture wastewater treatment, etc. Nullify all land use covenants that prohibit vegetable gardens.

11. Reintroduce and protect keystone species such as (in North America) beavers, wolves, and cougars.

12. Carry out water restoration projects worldwide through water retention landscapes (swales, ponds, check dams, etc.), regenerative grazing and horticulture, and the strategic removal of dams, canals, and levees.

13. Relocalize the food system and promote economic localization generally, first by nullifying free trade treaties and replacing them with "fair trade treaties" that protect local economic sovereignty.

14. Institute a negative-interest financial system through international agreement to impose liquidity fees on bank reserves, along with complementary measures such as Georgist land taxes and other anti-speculative taxes.

15. Apply pollution taxes to make companies internalize the social and ecological costs of toxic waste, radioactive waste, air pollution, and water pollution.

16. Impose a deposit system for most manufactured goods so that manufacturers have an incentive to create durable, repairable products with easily recoverable materials.

17. Turn away from pesticides.

In the conventional climate narrative, pesticides are virtually irrelevant to the fate of the biosphere. Not so in the living planet narrative.

In chapter 3 I referenced the ongoing insect holocaust, a term I do not apply lightly. From Europe to Australia to the Americas, insect biomass is in precipitous decline, a phenomenon that many scientists attribute to the growing use of insecticides over the past eighty years. Of particular concern are the notorious neonicotinoids, the most heavily used insecticides today. Because they are generally long-lived, these chemicals permeate the environment, showing up in plant nectar, pollen, ground-water, and soil.

Except in the case of honeybees and other pollinators, there is no direct proof that these are responsible for the insect holocaust, which in some areas approaches 90 percent (in terms of biomass decline). The scarcity of evidence is not surprising given that most research is funded by the same companies that manufacture the insecticides. Furthermore, current research methods are geared toward identifying monocausal phenomena, but insect decline is probably due to multiple, synergistic causes, including habitat disruption, soil degradation, drought, and other forms of chemical pollution. Yet for the insects, surely insecticides are the lynchpin.

Insects are a crucial part of nearly all food chains, as well as being instrumental in the life cycles of plants. Countless symbiotic relationships between insects and fungi, bacteria, worms, plants, and vertebrates maintain the web of life. Pesticides harm these other beings directly too, not only via their harm to insects. Aside from neonicotinoids, the other notorious pesticide today is the herbicide glyphosate, which also has pervasive ecological effects far beyond its time and place of application.

We have basically conducted an eighty-year experiment to see what happens to the biosphere when we constantly dump poison into it. Life is resilient, so the effects were hard to notice at first, but they have gathered now to critical mass.

To transition away from pesticides entails the wholesale deindustrialization of agriculture, in particular the ending of monocropping. The transition cannot happen overnight, but it needs to start now, and on a massive scale. What *can* happen overnight would be to ban completely the use of pesticides for nonagricultural use: lawn chemicals, garden insecticides, glyphosate used in city parks, etc. Aside from forest and wetlands destruction, pesticides might be the most urgent environmental issue we face. Insect decimation is no joke. Insects are life at nearly its most basic; they are fundamental tissues of the living planetary body. If we want a living planet (with, among other things, a healthy climate), we have to hear the message in the die-off and do something about it right now.

18. Demilitarize society.

As the saying goes, you cannot serve two masters. When a person or a society serves two conflicting aims, eventually the contradiction will rise to the surface in the form of a choice point, a crossroads, or a clarifying test.

What overarching aim does the military serve? Traditionally it was the interests of the nation-state; today it might be more the interests of transnational capital. On a deeper level, it serves the paradigm of domination through force. Demilitarization is therefore a necessary marker and sign of a civilizational shift of priorities. In a war, the top priority is the defeat of the enemy; anything else may need to be sacrificed. In a war, a country won't let environmental considerations interfere with bombing oil wells, pipelines, factories, and so forth. The air force won't limit its bombing runs to conserve fossil fuels. The army won't curtail its use of depleted uranium munitions for fear of contaminating groundwater. They are serving another master—something else comes first.

The environmental crisis invites us to change our priorities and put earth healing first, make ecological-social healing the primary conditioning parameter on every political decision. The military mind puts defeat of an enemy first instead. More tangibly, the military sucks up prodigious amounts of energy, materials, money, and human talent. Tens of thousands of the best scientists and engineers devote their lives to developing weaponry. Millions of healthy, capable, idealistic young people join the military. And of course, the money squandered on weaponry is enough to fund probably all of the other proposals in this book.

Representing an enormous commitment of human effort, the military is the concrete expression of an aim—well-being and progress through domination—that itself arises from the defining story of Separation. Demilitarization signals a profound shift in priorities as well as in the story underlying them.

Just as in personal life, psychological change requires concrete action in order to seem and be real. Demilitarization—closing down bases, retooling armaments factories, retraining troops, etc.—is a collective ritual demonstrating to the collective mind that everything is different now.

Rather than quantify the resources and energy that demilitarization would free up, I will just appeal to your intuition that we face a crossroads. War or peace? Love or fear? Domination or service? We are not going to see real earth healing while maintaining the military-industrial complex. If we want to live in a more beautiful world, we will have to give up central aspects of business-as-usual. What could be more starkly relevant a first step than demilitarization?

Note that I have not included a carbon tax among these proposals. The reasons are: (1) large reductions in fossil fuel use will necessarily result from establishing vast new marine sanctuaries and forest conservation zones, and from various pollution taxes and watershed restoration projects; (2) regenerative agriculture and reforestation will sequester large amounts of carbon; (3) carbon taxes create perverse incentives for things like large hydroelectric facilities and biofuels plantations that destroy ecosystems. This book has argued that while high levels of greenhouse gases add stress to an already challenged biosphere, the main problem is the impoverishment of life and the degradation of the water cycle. But even if I am wrong, the measures I have described will achieve carbon drawdown without making carbon the primary framing issue.

These measures are much more ambitious than merely shifting to carbon-neutral energy. I was about to say that they aren't going to "happen overnight," but let us not hold to that expectation too tightly. The process of change often includes long periods of apparent stasis, during which invisible substructures are shifting even as the visible superstructures look stronger and more permanent than ever. In fact, they are like a termite-ridden building that can indeed collapse overnight.

Even so, many of the changes make sense only from the ground of a new story. They will take time to germinate, blossom, and bear fruit. I applaud the urgency to act, but along with this urgency must come the patience to do things that take many generations to flower. We need to do some things that will bring quick results (many are listed above), but we also need to do things that will bring slow results. Which are yours to do? Are you excited to campaign for a plastic bag ban where you live? For a marine sanctuary? To stop a pipeline or fracking well? Or is your calling something that will take generations to have tangible ecological benefits? Is it maybe to work with trauma survivors? To aid refugees? To practice holistic midwifery? To be a mentor to at-risk youth? To raise children who carry a little less pain into adulthood than you did? These are the kind of things that enrich the cultural soil in which new paradigms and policies can grow. Furthermore, even though there is no clear, near-term causal through line from them to, say, watershed restoration or rainforest conservation, some part of me knows that they are indispensable. They are a declaration of the kind of world we want to live in; they are a prayer bringing us into alignment with a living world.

All of the policies and practices I have described are within reach right now. The vision of a Green World is not fantasy; nor, however, is it realistic. What it is, is possible. It requires each one of us to dedicate ourselves, unreasonably and with no guarantee of success, to our unique form of service. It requires that we trust our knowing that a healed world, a greened world, a more beautiful world is truly possible. I hope this book has amplified that calling and trued you to that possibility.

Bibliography

Ahmed, Shariqua. (2015). "How Rajendra Singh AKA 'Waterman of India' Solved Rural Rajasthan's Freshwater Crisis." www.dogonews.com/2015/10/22 /how-rajendra-singh-aka-waterman-of-india-solved-rural-rajasthans -freshwater-crisis.

Albert, Bruce, et al. (2014). "Rescuing US Biomedical Research from Its Systemic Flaws." *Proceedings of the National Academy of Sciences* 111, no. 16 (March 18).

Alley, R. B. (2000). "The Younger Dryas Cold Interval as Viewed from Central Greenland." *Quaternary Science Reviews* 19.

Anderson, Kat. (2006). Tending the Wild: Native American Knowledge and the Management of California's Natural Resources. Oakland, CA: University of California Press.

Andrich, M. A., and J. Imberger. (2013). "The Effect of Land Clearing on Rainfall and Fresh Water Resources in Western Australia: A Multi-functional Sustainability Analysis." *International Journal of Sustainable Development & World Ecology* 20, no. 6.

Angelini, I. M., M. Garstang, R. E. Davis, et al. (2011). "On the Coupling between Vegetation and the Atmosphere." *Theoretical and Applied Climatology* 105 (August): 243. doi:10.1007/s00704-010-0377-5.

Apfelbaum, Steve. (1993). "The Role of Landscapes in Stormwater Management." Applied Ecological Services. www.researchgate.net/publication/254840834 _The_Role_of_Landscapes_in_Stormwater_Management.

Apffel-Marglin, Frédérique. (2012). *Subversive Spiritualities.* New York: Oxford University Press.

Arneth, A., et al. (2017). "Historical Carbon Dioxide Emissions Caused by Land-Use Changes Are Possibly Larger Than Assumed." *Nature Geoscience* 10: 79–84.

Baccini, A., et al. (2012). "Estimated Carbon Dioxide Emissions from Tropical Deforestation Improved by Carbon-Density Maps." *Nature Climate Change* 2: 182–85. doi:10.1038/nclimate1354.

Baccini, A., et al. (2017). "Tropical Forests Are a Net Carbon Source based on Aboveground Measurements of Gain and Loss." *Science,* September 28: eaam5962. doi:10.1126/science.aam5962.

Baker, Monya. (2016). "1,500 Scientists Lift the Lid on Reproducibility." *Nature,* May 25.

Barnosky, Anthony D., et al. (2016). "Variable Impact of Late-Quaternary Megafaunal Extinction in Causing Ecological State Shifts in North and South America." *Proceedings of the National Academy of Sciences of the United States of America* 113, no. 4: 856–61.

Belluz, Julia, and Steven Hoffman. (2015). "Science Is Often Flawed: It's Time We Embraced That." *Vox,* May 13. www.vox.com/2015/5/13/8591837 /how-science-is-broken.

Biodiversity for a Livable Climate. (2017). *Compendium of Scientific and Practical Findings Supporting Eco-Restoration to Address Global Warming* 1, no. 1 (July). https://bio4climate.org/wp-content/uploads/Compendium-Vol-1-No-1-July -2017-Biodiversity-for-a-Livable-Climate-1.pdf.

Blocker, Jack. (2014). "Horrifying Machine Eats Entire Trees in Seconds." Metro News UK, June 28.

Bonan, G. B. (2008). "Forests and Climate Change: Forcings, Feedbacks, and the Climate Benefits of Forests." *Science* 320: 1444–49.

Buhner, Stephen Harrod. (2002). *The Lost Language of Plants.* White River Junction, VT: Chelsea Green.

Burtt, E. A. (1925). *The Metaphysical Foundations of Modern Science.* Reprinted by Dover Publications, 2003.

Carrington, Damian. (2016). "Global 'Greening' Has Slowed Rise of CO_2 in the Atmosphere, Study Finds." *The Guardian,* November 8.

Christensen, V., et al. (2014). "A Century of Fish Biomass Decline in the Ocean." *Marine Ecology Progress Series* 512, no. 1: 155–66.

Clay, Jason W., and Bonnie K. Holcomb. (1985). "The Politics of Famine in Ethiopia." *Cultural Survival,* June.

Community Cloud Forest Conservation. (2018). "Deforestation in Guatemala: Tracking by Decade." http://cloudforestconservation.org/knowledge/cloud-forest/deforestation/.

Cooperafloresta. (2016). "Pesquisas ajudam a comprovar benefícios das agroflorestas." *Divulgador de Notícias,* August 6.

Courcoux, Gaëlle. (2009). "Decline in Rainfall in the Amazon Basin." Institut de recherche pour le développement Scientific Newssheets, December. Translated by Nicholas Flay. https://en.ird.fr/the-media-centre/scientific-newssheets/336-decline-in-rainfall-in-the-amazon-basin.

Crist, Eileen. (2007). "Beyond the Climate Crisis: A Critique of Climate Change Discourse." *Telos* 4 (Winter): 29–55. www.umweltethik.at/wp/wp-content/uploads/CristBeyondTheClimateCrisis.pdf.

Crowther, T. W., et al. (2015). "Mapping Tree Density at a Global Scale." *Nature* 525 (September 10): 201–5.

Curry, Judith. (2016). "The Paradox of the Climate Change Consensus." *Climate Etc.,* April 17. https://judithcurry.com/2016/04/17/the-paradox-of-the-climate-change-consensus/#more-21437.

Daniels, Mitch. (2017). "Avoiding GMOs Is Not Only Unscientific, It Is Immoral." *Washington Post,* December 27.

Davidson, N. C. (2014). "How Much Wetland Has the World Lost? Long-Term and Recent Trends in Global Wetland Area." *Marine and Freshwater Research* 65, no. 10: 934–41. doi:10.1071/MF14173.

Dear, J., et al. (2013). "Life and Debt: Global Studies of Debt Resistance." Jubilee Debt Campaign, October.

Demenge, Jonathan. (2018). "Measuring Ecological Footprints of Subsistence Farmers in Ladakh." Institute of Development Studies.

DeRamus, H. A., et al. (2003). "Methane Emissions of Beef Cattle on Forages: Efficiency of Grazing Management Systems." *Journal of Environmental Quality* 32, no. 1 (January–February). doi:10.2134/jeq2003.2690.

Desmond, Matthew. (2017). *Evicted: Poverty and Profit in the American City.* New York: Broadway Books.

Dewar, W. K., et al. (2006). "Does the Marine Biosphere Mix the Ocean?" *Journal of Marine Research* 64: 541–61.

Diamond, Jared. (2005). Collapse: How Societies Choose to Fail or Succeed. New York: Viking Press.

Doughty, C. E., et al. (2016). "Global Nutrient Transport in a World of Giants." *Proceedings of the National Academy of Sciences* 113, no. 4 (January 26): 868–73.

Doughty, Christopher E., Adam Wolf, and Yadvinder Malhi. (2013). "The Legacy of the Pleistocene Megafauna Extinction on Nutrient Availability in Amazonia." *Nature Geoscience* 6: 761–64. doi:10.1038/ngeo1895.

Duarte, C. M., T. Sintes, and N. Marbà. (2013). "Assessing the CO_2 Capture Potential of Seagrass Restoration Projects." *Journal of Applied Ecology* 50: 1341–49. doi:10.1111/1365-2664.12155.

The Economist. (2013). "Trouble at the Lab." *The Economist,* October 18.

Eisenstein, Charles. (2014). "The Waters of Heterodoxy." November 1. https://charleseisenstein.net/essays/the-waters-of-heterodoxy-g-pollacks-the-fourth-phase-of-water/.

———. (2015a). "*Aluna:* A Message to Little Brother." *Tikkun,* May 26.

———. (2015b). "'Don't Owe. Won't Pay.' Everything You've Been Told about Debt Is Wrong." *YES!* magazine, August 20.

———. (2018). "Opposition to GMOs Is Neither Unscientific Nor Immoral." January 9. https://charleseisenstein.net/essays/opposition-to-gmos/.

Ellison, D., et al. (2017). "Trees, Forests and Water: Cool Insights for a Hot World." *Global Environmental Change* 43.

Fears, Darryl. (2013). "Study Says U.S. Can't Keep Up with Loss of Wetland." *Washington Post,* December 8.

Ferroni, Ferruccio, and Robert J. Hopkirk. (2016). "Energy Return on Energy Invested (ERoEI) for Photovoltaic Solar Systems in Regions of Moderate Insolation." *Energy Policy* 94: 336–44.

Food and Agriculture Organization of the United Nations. (2009). "Pastoralists—Playing a Critical Role in Managing Grasslands for Climate Change Mitigation and Adaptation." www.fao.org/fileadmin/templates/agphome/documents/climate/Grasslands_Brief_final.pdf.

———. (2010). Global Forest Resources Assessment 2010. FAO Forestry Paper 163.

Foster, Grant. (2016). "Which Satellite Data?" *Open Mind,* November 27. https://tamino.wordpress.com/2016/11/27/which-satellite-data/.

Freedman, David H. (2010). "Lies, Damned Lies, and Medical Science." *The Atlantic.*

Gordon, Robert. (2012). "Is U.S. Economic Growth Over? Faltering Innovation Confronts the Six Headwinds." NBER Working Paper No. 18315. National Bureau of Economic Research, August. doi:10.3386/w18315.

Gorshkov, V. G., and A. M. Makarieva. (2006). "Biotic Pump of Atmospheric Moisture as Driver of the Hydrological Cycle on Land." *Hydrology and Earth System Sciences Discussions* 3.

Hallmann, C. A., et al. (2017). "More Than 75 Percent Decline over 27 Years in Total Flying Insect Biomass in Protected Areas." *PLoS ONE* 12, no. 10: e0185809. doi:10.1371/journal.pone.0185809.

Hance, Jeremy. (2012). "New Meteorological Theory Argues That the World's Forests Are Rainmakers." *Mongabay,* February 1. https://news.mongabay.com/2012/02 /new-meteorological-theory-argues-that-the-worlds-forests-are-rainmakers/.

Harball, Elizabeth. (2014). "How Fish Cool Off Global Warming." *Scientific American,* June 9.

Hausfather, Zeke, and Matthew Menn. (2013). "Urban Heat Islands and U.S. Temperature Trends." *RealClimate,* February 13.

Hawken, Paul. (2017). *Drawdown.* New York: Penguin Books.

Hesslerová, P., J. Pokorný, J. Brom, and A. Rejšková–Procházková. (2013). "Daily Dynamics of Radiation Surface Temperature of Different Land Cover Types in a Temperate Cultural Landscape: Consequences for the Local Climate." *Ecological Engineering* 54: 145–54. doi:10.1016/j.ecoleng.2013.01.036.

Hoegh-Guldberg, Ove, et al. (2015). "Reviving the Ocean Economy: The Case for Action—2015." WWF International, April 22. www.worldwildlife.org /publications/reviving-the-oceans-economy-the-case-for-action-2015.

Horton, Scott. (2010). "Churchill's Dark Side: Six Questions for Madhusree Mukerjee." *Harper's Magazine,* November 4. https://harpers.org/blog/2010/11 /churchills-dark-side-six-questions-for-madhusree-mukerjee/.

Hughes, J. Donald. (2014). Environmental Problems of the Greeks and Romans: Ecology in the Ancient Mediterranean. Baltimore: Johns Hopkins University Press.

Hunt, Terry. (2006). "Rethinking the Fall of Easter Island." *American Scientist,* September–October.

Hunziker, Robert. (2018). "Insect Decimation Upstages Global Warming." *Counterpunch,* March 27. www.counterpunch.org/2018/03/27/insect-decimation-upstages-global-warming/.

Jehne, Walter. (2007). "The Biology of Global Warming and Its Profitable Mitigation." *Nature and Society,* December 2006–January 2007: 7–14.

Jenkinson, Stephen. (2018). *Come of Age: The Case for Elderhood in a Time of Trouble.* Berkeley, CA: North Atlantic Books, 2018.

Ko, Lisa. (2016). "Unwanted Sterilization and Eugenics Programs in the United States." PBS *Independent Lens,* January 29. www.pbs.org/independentlens/blog/unwanted-sterilization-and-eugenics-programs-in-the-united-states/.

Kopenawa, Davi, and Bruce Albert. (2013). *The Falling Sky: Words of a Yanomami Shaman.* Cambridge, MA: Harvard University Press.

Koppelaar, R. H. E. M. (2017). "Solar-PV Energy Payback and Net Energy: Meta-assessment of Study Quality, Reproducibility, and Results Harmonization." *Renewable and Sustainable Energy Reviews* 72 (May): 1241–55.

Kravčík, M., et al. (2007). *Water for the Recovery of the Climate—A New Water Paradigm.* Translated by David McLean and Jonathan Gresty. www.waterparadigm.org/download/Water_for_the_Recovery_of_the_Climate_A_New_Water_Paradigm.pdf.

Krüger, Michael. (2013). "The Rise and Fall of the Hockey Stick Charts." Translated, condensed, and edited by P. Gosselin. *Science Skeptical Blog.* http://notrickszone.com/2013/10/17/climatology-sees-one-of-the-greatest-scientific-reversals-of-all-time-the-rise-and-fall-of-the-hockey-stick-charts/#sthash.Tcrq2TzK.dpuf.

Kwok, Roberta. (2009). "Fish Are Crucial in Oceanic Carbon Cycle." *Nature,* January 15.

Lawrence, Jane. (2000). "The Indian Health Service and the Sterilization of Native American Women." *American Indian Quarterly* 24, no. 3 (Summer): 400–419. www.jstor.org/stable/1185911.

Life in Syntropy. (2015). "Life in Syntropy." Film. https://lifeinsyntropy.org.

Light, Malcolm. (2014). "Focus on Methane." *Arctic News,* July 14. https://arctic-news.blogspot.ca/2014/07/focus-on-methane.htm.

Lovins, L. Hunter. (2014). "Why George Monbiot Is Wrong: Grazing Livestock Can Save the World." *The Guardian,* August 19.

Luoma, Jon. (2012). "China's Reforestation Program: Big Success or Just an Illusion?" *Yale Environment 360,* January 17. https://e360.yale.edu/features/chinas_reforestation_programs_big_success_or_just_an_illusion.

Machmuller, Megan B., Mark G. Kramer, Taylor K. Cyle, Nick Hill, and Dennis Hancock. (2015). "Emerging Land Use Practices Rapidly Increase Soil Organic Matter." *Nature Communications* 6, article no. 6995. doi:10.1038/ncomms7995.

MacKinnon, J. B. (2013). *The Once and Future World.* Boston: Houghton Mifflin Harcourt.

Magill, Bobby. (2014). "Methane Emissions May Swell from behind Dams." Climate Central, October 29. www.scientificamerican.com/article/methane-emissions-may-swell-from-behind-dams/#.

Mahowald, Natalie M., et al. (2017). "Are the Impacts of Land Use on Warming Underestimated in Climate Policy?" *Environmental Research Letters* 12, no. 9 (September 18).

Marinelli, Janet. (2017). "In the Sierras: New Approaches to Protecting Forests Under Duress." *Yale Environment 360,* February 13. https://e360.yale.edu/features/in-the-sierras-new-thinking-on-protecting-forests-under-stress.

McNeil, Ben. (2014). "Is There a Creativity Deficit in Science?" *The New Atlantis,* September 3.

Middleton, David. (2012). "A Brief History of Atmospheric Carbon Dioxide Record Breaking." *Watts Up With That?* December 7. https://wattsupwiththat.com/2012/12/07/a-brief-history-of-atmospheric-carbon-dioxide-record-breaking/.

Millán, M. M. (2014). "Extreme Hydrometeorological Events and Climate Change Predictions in Europe." *Journal of Hydrology* 518: 206–24. doi:10.1016/j.jhydrol.2013.12.041.

Monbiot, George. (2008). "Small Is Bountiful." www.monbiot.com/2008/06/10/small-is-bountiful/.

Mongabay. (2018). "Environmental Profile" pages. https://rainforests.mongabay.com/countries.htm.

Moreno, C., D. S. Chassé, and L. Fuhr. (2015). "Carbon Metrics: Global Abstractions and Ecological Epistemicide." *Heinrich Böll Stiftung Publication Series on Ecology* 42.

Moriarty, Tom. (2010). "Tree Rings: Proxies for Temperature or CO2?" https://climatesanity.wordpress.com/2010/02/15/tree-rings-proxies-for-temperature-or-co2/.

Mothincarnate. (2015). "Does Urban Heat Island Effect Exaggerate Global Warming Trends?" Skeptical Science. https://skepticalscience.com/urban-heat-island-effect.htm.

Muller, Richard. (2004). "Global Warming Bombshell." *MIT Technology Review.*

Nellemann, C., et al., eds. (2009). "Blue Carbon: A Rapid Response Assessment." United Nations Environment Programme, GRID Arendal. www.grida.no.

The New Atlantis. (2006). "Rethinking Peer Review." *The New Atlantis,* no. 13 (Summer): 106–10.

Nicholls, Steve. (2009). *Paradise Found: Nature in America at the Time of Discovery.* Chicago: University of Chicago Press.

NOAA Geophysical Fluid Dynamics Laboratory. (2018). "Global Warming and Hurricanes: An Overview of Current Research Results." January 24. www.gfdl.noaa .gov/global-warming-and-hurricanes/.

Noble, Denis. (2017). *Dance to the Tune of Life.* New York: Cambridge University Press.

Ohlson, Kristin. (2014). *The Soil Will Save Us.* Harlan, IA: Rodale Books.

Orion, Tao. (2015). *Beyond the War on Invasive Species.* White River Junction, VT: Chelsea Green.

Pan, Y., et al. (2011). "A Large and Persistent Carbon Sink in the World's Forests." *Science* 333, no. 6045 (August 19): 988–93. doi:10.1126/science.1201609.

Pearce, Fred. (2017). "How Big Water Projects Helped Trigger Africa's Migrant Crisis." *Yale Environment 360,* October 17.

Peplow, Mark. (2014). "Social Sciences Suffer from Severe Publication Bias." *Nature,* August 28.

Prashad, Vijay. (2017). "The Human Carnage from Billionaires Trying to Carve Up the Planet to Build Their Empires Is Astounding." *Alternet,* August 16. www .alternet.org/world/human-carnage-billionaires-trying-carve-planet-build-their -empires-astounding.

Ridley, Matt. (2015). "The Climate Wars' Damage to Science." *Quadrant Online,* June 19.

Robbins, Jim. (2017). "Why the World's Rivers Are Losing Sediment and Why It Matters." *Yale Environment 360,* June 20.

Robertson, Joshua. (2017). "'Alarming' Rise in Queensland Tree Clearing as 400,000 Hectares Stripped." *The Guardian,* October 5.

Rodale Institute. (2014). "Regenerative Organic Agriculture and Climate Change." Rodale Institute White Paper, April 17. http://rodaleinstitute.org /regenerative-organic-agriculture-and-climate-change/.

Roman, Joe, and Stephen R. Palumbi. (2003). "Whales before Whaling in the North Atlantic." *Science* 301, no. 5632 (July 25): 508–10.

Rosa, Isabel M. D., et al. (2016). "The Environmental Legacy of Modern Tropical Deforestation." *Current Biology* 26, no. 16 (August 22): 2161–66.

Ruddiman, William. (2003). "The Anthropogenic Greenhouse Era Began Thousands of Years Ago." *Climatic Change* 61: 261–93.

Runyan, C., and Paolo D'Odorico. (2016). *Global Deforestation.* New York: Cambridge University Press.

Russian Federation, Federal State Statistics Service. (2018). "Agricultural Production by Types of Enterprise (Percent)." www.gks.ru/wps/wcm/connect/rosstat_main /rosstat/en/figures/agriculture/.

Sabajo, C. R., et al. (2017). "Expansion of Oil Palm and Other Cash Crops Causes an Increase of the Land Surface Temperature in the Jambi Province in Indonesia." *Biogeosciences* 14: 4619–35. doi:10.5194/bg-14-4619-2017.

Sachs, Wolfgang, ed. (2010). The Development Dictionary: A Guide to Knowledge as Power. London: Zed Publishing.

Savory, Alan. (2013). "How to Fight Desertification and Reverse Climate Change." *TED2013.* www.ted.com/talks/allan_savory_how_to_green_the_world _s_deserts_and_reverse_climate_change.

Schellnhuber, Hans-Joachim. (2004). *Earth System Analysis for Sustainability.* Cambridge, MA: MIT Press.

Schiermeier, Quirin. (2008). "'Rain-making' Bacteria Found around the World." *Nature,* February 28. doi:10.1038/news.2008.632.

Schiffman, Richard. (2015). "How Can We Make People Care about Climate Change?" *Yale Environment 360,* July 9.

Schwartz, Judith. (2013). "Clearing Forests May Transform Local—and Global— Climate." *Scientific American,* March 4.

———. (2016). *Water in Plain Sight.* New York: St. Martin's Press.

Sendin, Patricia. (2016). "Syntropic Agriculture: The Regenerative Food-Growing Method That Could Reverse Climate Change and End Hunger." *Not Only about Architecture,* August 12. www.patriciasendin.com/2016/08/syntropic-agriculture -regenerative-food.html.

Shapiro, James. (2011). *Evolution: A View from the 21st Century.* Upper Saddle River, NJ: Prentice Hall.

Sharaskin, Leonid. (2008). "The Socioeconomic and Cultural Significance of Food Gardening in the Vladimir Region of Russia." University of Missouri, Columbia. http://naturalhomes.org/naturalliving/russian-dacha.htm.

Sierra Forest Legacy. (2012). "Logging Impacts." www.sierraforestlegacy.org /FC_FireForestEcology/FFE_LoggingImpacts.php.

Smith, Richard. (2006). "Peer Review: A Flawed Process at the Heart of Science and Journals." *Journal of the Royal Society of Medicine* 99, no. 4 (April): 178–82.

Smith, Vincent H. (2016). "Crony Farmers: Farm Subsidies Exist Because of Political Power, Not Economics." *US News and World Report,* January 14. www.usnews .com/opinion/economic-intelligence/articles/2016-01-14/farm-subsidies-are -crony-capitalism.

Soga, M., et al. (2017). "Gardening Is Beneficial for Health: A Meta-analysis." *Preventive Medicine Reports* 5: 92–99. doi:10.1016/j.pmedr.2016.11.007.

Spencer, Roy. (2016). "Comments on New RSS v4 Pause-Busting Global Temperature Dataset." March 4. www.drroyspencer.com/2016/03/comments-on-new -rss-v4-pause-busting-global-temperature-dataset/.

Spielmaker, D. M. (2018). Growing a Nation Historical Timeline. March 21. www.agclassroom.org/gan/timeline/index.htm.

Spracklen, Dominick V., et al. (2008). "Boreal Forests, Aerosols and the Impacts on Clouds and Climate." *Philosophical Transactions of the Royal Society A,* December 28. doi:10.1098/rsta.2008.0201.

Steele, Jim. (2013). "Unwarranted Temperature Adjustments: Conspiracy or Ignorance?" *Landscapes and Cycles.*

Stoknes, Per Espen. (2015). *What We Think about When We Try Not to Think about Global Warming.* White River Junction, VT: Chelsea Green.

Taguchi, Viviane. (2016). "Agricultura Sintrópica, SP." *Globo Rural* no. 370 (August). Editora Globo.

Teuling, Adriaan, et al. (2017). "Observational Evidence for Cloud Cover Enhancement over Western European Forests." *Nature Communications* 8 (January 11). doi:10.1038/ncomms14065.

Thompson, Andrea. (2008). "Earth's Clouds Alive with Bacteria." *Live Science,* February 27. www.livescience.com/2333-earth-clouds-alive-bacteria.html.

Trenberth, Kevin E., and David P. Stepaniak. (2004). "The Flow of Energy through the Earth's Climate System." *Quarterly Journal of the Royal Meteorological Society* 130: 2677–701. doi:10.1256/qj.04.83.

Turner, Scott J. (2017). *Purpose and Desire.* New York: HarperOne.

Ünal, Fatma Gül. (2008). "Small Is Beautiful: Evidence of Inverse Size Yield Relationship in Rural Turkey." Levy Economics Institute Working Paper No. 551 (December 5).

Van Den Berg, J., and P. Rietveld. (2004). "Reconsidering the Limits to World Population: Meta-analysis and Meta-prediction." *BioScience* 54, no. 3 (March 1): 195–204.

Walton, Alice. (2015). "Why the Super-Successful Get Depressed." *Forbes,* January 26.

Watts, Anthony. (2009). "Is the U.S. Surface Temperature Record Reliable?" Chicago: Heartland Institute.

Waycott, Michelle, et al. (2009). "Accelerating Loss of Seagrasses across the Globe Threatens Coastal Ecosystems." *Proceedings of the National Academy of Sciences of the United States of America* 106, no. 30: 12377–81.

Weisman, Alan. (2008). "Africa after Us: What Effects Have Human Actions Had on the Sahara—The World's Largest Non-polar Desert?" *The Globalist,* January 26.

Weisse, Mikaela, and Liz Goldman. (2017). "Global Tree Cover Loss Rose 51% in 2016." Global Forest Watch, October 18. https://blog.globalforestwatch.org /data/global-tree-cover-loss-rose-51-percent-in-2016.html.

Whitfield, John. (2003). "Whaling Blamed for Seal and Otter Slumps." *Nature,* September 23. www.nature.com/news/2003/030922/full/news030922-5.html.

The World Factbook of the Central Intelligence Agency. "Field Listing: Total Fertility Rate." www.cia.gov/library/publications/the-world-factbook/fields/2127.html.

World Wildlife Federation. (2015). Living Blue Planet Report: Species, Habitats, and Human Wellbeing.

Wuerthner, George. (2016). "The Myth That Logging Prevents Forest Fires." Counterpunch, April 19. www.counterpunch.org/2016/04/19/the-myth-that-logging -prevents-forest-fires/.

Yang, Xiaoping, et al. (2015). "Groundwater Sapping as the Cause of Irreversible Desertification of Hunshandake Sandy Lands, Inner Mongolia, Northern China." *Proceedings of the National Academy of Sciences of the United States of America.* doi:10.1073/pnas.1418090112.

Yirka, Bob. (2015). "Study Indicates Groundwater Sapping Led to Desertification of Parts of Inner Mongolia." *Phys.org,* January 6. https://phys.org/news/2015-01 -groundwater-sapping-desertification-mongolia.html.

Zubrin, Robert. (2012). "The Population Control Holocaust." *The New Atlantis* 35 (Spring): 33–54.

Index

Index

About the Author

CHARLES EISENSTEIN is a speaker and writer focusing on themes of civilization, consciousness, money, and human cultural evolution. His viral short films and essays online have established him as a genre-defying social philosopher and countercultural intellectual. Eisenstein graduated from Yale University in 1989 with a degree in mathematics and philosophy and spent the next ten years as a Chinese-English translator. He was interviewed by Oprah Winfrey on *SuperSoul Sunday*, July 16, 2017. He is the author of *The More Beautiful World Our Hearts Know Is Possible* (North Atlantic Books, 2013), *Sacred Economics* (North Atlantic Books, 2011), and *The Ascent of Humanity* (North Atlantic Books, 2013).

ALSO BY CHARLES EISENSTEIN

available from North Atlantic Books

*The More Beautiful World
Our Hearts Know Is Possible*
978-1-58394-724-1

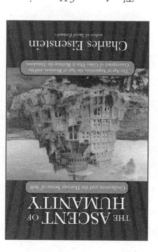

The Ascent of Humanity
978-1-58394-636-7 (hardcover)
978-1-58394-535-3 (paperback)

Sacred Economics
978-1-58394-397-7

North Atlantic Books
www.northatlanticbooks.com

North Atlantic Books is an independent, nonprofit publisher committed to a bold exploration of the relationships between mind, body, spirit, and nature.

About North Atlantic Books

North Atlantic Books (NAB) is an independent, nonprofit publisher committed to a bold exploration of the relationships between mind, body, spirit, and nature. Founded in 1974, NAB aims to nurture a holistic view of the arts, sciences, humanities, and healing. To make a donation or to learn more about our books, authors, events, and newsletter, please visit www.northatlanticbooks.com.